**BASIC operational amplifiers**

**Butterworths BASIC Series** includes the following titles:

BASIC aerodynamics
BASIC artificial intelligence
BASIC economics
BASIC hydraulics
BASIC hydrology
BASIC interactive graphics
BASIC investment appraisal
BASIC materials studies
BASIC matrix methods
BASIC mechanical vibrations
BASIC molecular spectroscopy
BASIC numerical mathematics
BASIC operational amplifiers
BASIC soil mechanics
BASIC statistics
BASIC stress analysis
BASIC theory of structures
BASIC thermodynamics and heat transfer

# BASIC operational amplifiers

**J C C Nelson,** BSc, PhD, CEng, MIEE
Lecturer
Department of Electrical and Electronic Engineering
University of Leeds

**Butterworths**
London . Boston . Durban . Singapore . Sydney . Toronto . Wellington

All rights reserved. No part of this publication may be reproduced or transmitted in any form or by any means, including photocopying and recording, without the written permission of the copyright holder, application for which should be addressed to the Publishers. Such written permission must also be obtained before any part of this publication is stored in a retrieval system of any nature.

This book is sold subject to the Standard Conditions of Sale of Net Books and may not be resold in the UK below the net price given by the Publishers in their current price list.

First published 1986

© Butterworth & Co. (Publishers) Ltd. 1986

---

**British Library Cataloguing in Publication Data**

Nelson, J.C.C.
   BASIC operational amplifiers.—
(Butterworths BASIC series)
1. Operational amplifiers
I. Title
621.3815′35        TK7871.58.06

ISBN 0-408-01580-2

---

**Library of Congress Cataloging in Publication Data**

Nelson, J. C. C. (John Christopher Cunliffe),
  1938–
BASIC operational amplifiers.

(Butterworths BASIC series)
Includes index.
1. Operational amplifiers—Data processing.
2. BASIC (Computer program language)  I. Title.
II. Series.
TK7871.58.06N45   1986     621.3815′35     86-12969
ISBN 0-408-01580-2

---

Photoset by Mid-County Press, 2a Merivale Road, London SW15 2NW
Printed and bound in England by Page Bros Ltd, Norwich, Norfolk

# Preface

The operational amplifier is essentially an electronic circuit capable of producing an output which is related to its input by a known mathematical operation. Originally such circuits were cumbersome and expensive since they made use of several thermionic valves and, subsequently, discrete transistors. Today 'op amps', as they have become popularly known, are available as integrated circuit 'chips' at very low cost. Four can be accommodated in one small package at a cost per amplifier of 10p or less. Consequently, they are used in a remarkably wide range of applications, not all of which are directly related to the original intention of performing mathematical operations. Most of the important application areas are discussed in this book.

All electronic circuit design involves substantial calculation in order to meet the required specification. In many cases the design becomes a compromise which can be achieved only after several initial 'guesses' at important parameters. Historically, the tedious repetitive calculations had to be performed manually, except by the privileged few with access to a computer. The development of microprocessors and the emergence of home and personal computers means that simple computer-aided design techniques are now available to everyone.

In this book, simple BASIC programs are related directly to the introductory text on operational amplifiers. The reader, with access to a home computer can obtain a 'feel' for the variables involved at each stage. He or she is encouraged to try a wide range of different parameter values, including 'silly' ones, in order to 'see what happens'.

The programs use only the commands that are available on almost all implementations of BASIC and should, therefore, run on any machine which offers this language. Machine-specific features such as graphics and colour have been deliberately avoided. This means that the output of programs is unspectacular and not particularly well formatted. However, there is tremendous scope for the keen reader who wishes to embellish the programs to suit a particular machine.

For example, colour could be used with red for unsatisfactory values and green for satisfactory ones.

All the programs given have been thoroughly tested. However, it is inevitable that there will be some sets of conditions under which programs may fail or give a misleading result. The author would be pleased to hear of any such 'bugs'.

The book assumes a background in the basic techniques of circuit theory, particularly the use of **j** notation for reactive circuits, with a corresponding level of mathematical ability. The Laplace transform is used in the chapter on active filters but not elsewhere. Practical considerations in the use of operational amplifiers are not discussed in detail; for this the reader is referred to a text such as *Operational amplifiers and linear integrated circuits* by Coughlin and Driscoll (Prentice Hall, 1982).

JCCN

# Contents

Preface

**1 Introduction to BASIC**    1

PROGRAM
1.1 Demonstration program    2

**2 Introduction to operational amplifier circuits**    5

2.1 The basic amplifier    5
2.2 Inverting mode, operation as scaler and summer    6
2.3 Non-inverting mode, voltage follower    11
2.4 Differential mode    14
2.5 Common mode rejection    16
2.6 Instrumentation amplifier    19
2.7 Exercises    22

PROGRAMS
2.1 Input and feedback resistances for operational amplifiers    10
2.2 Input and feedback resistances for operational amplifiers used in the non-inverting configuration    15
2.3 Determination of common mode rejection ratio    19
2.4 Resistance value for instrumentation amplifier    21

**3 Frequency response**    24

3.1 Open loop behaviour, compensation    24
3.2 Closed loop response, rise time    28
3.3 Large signal operation, slew rate and full power bandwidth    29
3.4 Choice of amplifier to meet a given specification    32
3.5 Exercises    36
3.6 Reference    36

### PROGRAMS
3.1 Full power bandwidth and slew rate conversion   32
3.2 Most suitable amplifier for a specified gain and bandwidth   34

## 4 Offset errors   37

4.1 Offset voltage, bias and difference currents   37
4.2 Temperature and other effects   40
4.3 Use of T network to reduce feedback resistance   42
4.4 Blocking of D.C. offset   44
4.5 Exercises   45
4.6 Reference   46

### PROGRAMS
4.1 Determination of offset errors   39
4.2 Worst case drift performance   41
4.3 T-network feedback component values   44

## 5 Waveform generation   47

5.1 Preliminary comments   47
5.2 Ramp-based generators   47
5.3 Sine wave oscillators   51
5.4 Exercises   55
5.5 References   55

### PROGRAMS
5.1 Component value determination for triangular wave oscillator   51
5.2 Component value determination for Wien bridge oscillator   54

## 6 Introduction to active filters   57

6.1 Preliminary comments   57
6.2 First order active filters   60
6.3 Second order active filters   63
6.4 State variable filters   77
6.5 Band rejection filters   79
6.6 Exercises   82
6.7 References   83

### PROGRAMS
6.1 Order of filter required for a specified rejection   59
6.2 Sallen–Key unity gain filter design   67
6.3 Sallen–Key equal component filter design   71

|     |                                       |     |
| --- | ------------------------------------- | --- |
| 6.4 | Bandpass filter design                | 75  |
| 6.5 | Notch filter design                   | 81  |

## 7 Non-linear circuits       84

|     |                                       |     |
| --- | ------------------------------------- | --- |
| 7.1 | Preliminary comments                  | 84  |
| 7.2 | Simple limiting                       | 84  |
| 7.3 | Precision limiting and rectification  | 89  |
| 7.4 | Arbitrary function generators         | 95  |
| 7.5 | Logarithmic amplifiers                | 103 |
| 7.6 | Exercises                             | 105 |
| 7.7 | References                            | 106 |

PROGRAMS

|     |                                            |     |
| --- | ------------------------------------------ | --- |
| 7.1 | Feedback limiter design                    | 88  |
| 7.2 | Precision limiter design                   | 92  |
| 7.3 | Simple diode function generator design     | 98  |

Appendix: Preferred component values     107

PROGRAM

A.1 Nearest preferred value     108

Index     111

Chapter 1
# Introduction to BASIC

The purpose of this chapter is to assist those readers with no previous knowledge of programming in BASIC; others may safely turn to Chapter 2. It does not in any way replace the manual for the particular computer being used; this should always be available for consultation.

By way of illustration, a simple program, P(1.1), will be presented. This involves most of the statements used in the programs in this book. Functions will not be specifically considered here since those required are directly related to conventional mathematical notation. The purpose and operation of each statement will be described in detail.

Lines 10 to 40 are purely comment (or REMark in BASIC). The computer ignores the rest of the line following REM. However, these lines are very important since they allow the title and purpose of the program to be specified (in this case the conversion of angular frequencies to hertz). Remark statements should also be used elsewhere in the program if required in order to explain operations whose purpose may not be clear otherwise.

In BASIC, PRINT statements normally relate to the display screen; special commands are required to send data to a printer, if one is available. The specimen runs presented throughout this book were obtained by modifying the programs for this purpose; as listed they provide output direct to the screen. The heart-shaped symbol in line 50 is the Commodore symbol for a command which clears the screen in preparation for a program run. In many versions of BASIC this is a 'clear screen' (CLS) command; consult your manual if in doubt.

The frequencies before and after conversion are to be stored in 'arrays'. Computers differ in their requirements for specifying the required maximum array sizes in advance. It is good practice to do so in any case, by means of a DIMension statement as in line 60. Notice that line 60 actually contains two statements, which must be separated by a colon. Most realizations of BASIC accept this but you should check to see how many statements are permitted.

Line 70 asks how many frequencies are to be converted and line 80

1

## 2  Introduction to BASIC

```
10 REM PROGRAM P(1.1)
20 REM DEMONSTRATION PROGRAM
30 REM THIS PROGRAM CONVERTS ANGULAR
40 REM FREQUENCIES TO HZ:
50 PRINT ""
60 DIM W(10):DIM F(10)
70 PRINT "NUMBER OF FREQUENCIES TO BE CONVERTED"
80 INPUT N
90 FOR I = 1 TO N
100 PRINT "ANGULAR FREQUENCY #"I"(RADIANS PER SECOND)"
110 INPUT W(I)
120 NEXT I
130 PRINT "ANGULAR FREQUENCY    HZ"
140 FOR I = 1 TO N
150 LET F(I) = W(I)/(2*π)
160 PRINTTAB(7) W(I),F(I)
170 NEXT I
180 PRINT "ANY MORE FREQUENCIES? (Y/N)"
190 GET A$: IF A$ = "" THEN GOTO 190
200 IF A$ = "Y" THEN 70
210 END
```

*Specimen run*

```
NUMBER OF FREQUENCIES TO BE CONVERTED
? 5
ANGULAR FREQUENCY # 1 (RADIANS PER SECOND)
? 10
ANGULAR FREQUENCY # 2 (RADIANS PER SECOND)
? 500
ANGULAR FREQUENCY # 3 (RADIANS PER SECOND)
? 10000
ANGULAR FREQUENCY # 4 (RADIANS PER SECOND)
? .15
ANGULAR FREQUENCY # 5 (RADIANS PER SECOND)
? 400
ANGULAR FREQUENCY    HZ
       10            1.59154943
      500           79.5774716
    10000         1591.54943
      .15            .0238732415
      400           63.6619773
ANY MORE FREQUENCIES? (Y/N)
NUMBER OF FREQUENCIES TO BE CONVERTED
? 1
ANGULAR FREQUENCY # 1 (RADIANS PER SECOND)
? 80
ANGULAR FREQUENCY    HZ
       80           12.7323955
ANY MORE FREQUENCIES? (Y/N)
```

prints a question mark and then waits for the required number to be entered. This number *must* be followed by pressing the RETURN key since the computer has no way of knowing how many digits to expect in the input. It is often possible to combine these two lines into an

INPUT statement which also contains the required message. It is assumed that the number of frequencies to be converted will be less than 10, as declared in line 60. If this is not the case an error will occur later when the array overflows.

When several similar operations are to be performed, these can conveniently be specified by means of a 'loop', which provides the required repetition. A simple example occurs in lines 90 to 120; the loop starts with FOR and ends with NEXT. The value of I will initially be set to 1 (as specified in line 90) and will be increased by 1 each time the NEXT statement is encountered. This will be followed by a return to line 90, unless I exceeds N in which case the program will advance to line 130 and leave the loop. The statements inside the loop (lines 100 and 110) will, therefore, be executed N times. These statements ask for, and accept, each of the required angular frequencies. Notice that information within quotation marks is displayed by a PRINT statement without modification. However, the I in line 100 is outside the quotation marks and the current value of I will be printed (see the specimen run of the program).

Line 130 merely prints the appropriate headings for the output that is to follow. Lines 140 to 170 form another FOR loop in which the required values are calculated and the output printed. In line 150 the current value of angular frequency (W(I)) is divided by $2 \times \pi$ in order to give the required frequency in Hz (F(I)). The 'LET' in this statement is optional in almost all realizations of BASIC and is normally omitted. It is important to realize that statements of this kind are *not* equations in the mathematical sense. For example, N = N + 1 would mean that the stored value of N is to be increased by 1, whereas mathematically it is meaningless. 'TAB' in line 160 ensures that printing is offset (by 7 in this case) so that the values appear under the previously printed headings. The comma between W(I) and F(I) inserts a reasonable space between the two values.

A simple question is asked in line 180 and is followed by a GET statement. This differs from an INPUT statement in that only one character, without operation of RETURN, is expected. (In some realizations this command is known as INKEY; check your manual for details.) As the system does not wait for RETURN the program would advance to the next line before the operator has chance to press a key. The second statement in line 190 overcomes this difficulty. The two quotation marks with nothing between them represent a null character, which will be the value read by the computer when no key is pressed. As long as this condition persists, the IF ... THEN statement will hold the program in line 190. As soon as any key is pressed, the program will advance to line 200. If that key is Y, line 200 will cause a jump back to line 70 and more frequencies can be converted. Any other key causes line 210 to be entered, which terminates the program. A variable name followed by a dollar sign ($) represents a 'string' (lines 190 and 200). A string variable can consist of alphanumeric characters and other symbols whereas plain variables are expected by the computer to be purely numerical. String variables in program statements must be enclosed in quotation marks (line 200); if they were absent line 200 would produce an error since A$ and Y are variables of different types.

Notice that GOTO statements may be used on their own to provide an unconditional jump (although they are not favoured in good programming practice). In conditional jumps (IF ... THEN) the GOTO is usually optional. It is included in line 190 for completeness but omitted in line 200.

The specimen run, obtained as mentioned above, shows a typical sequence which would be displayed on the screen after RUN is typed and the RETURN key pressed. The machine responds by clearing the screen and printing 'Number of frequencies to be converted'. The question mark on the next line is the prompt produced by the input statement (line 80) but 5 is the value (followed by RETURN) typed in by the operator.

The five angular frequencies are obtained in an exactly similar way after which the table is displayed without further intervention (the program 'knows' that five frequencies will be entered from the value of N).

The operator is then asked if he wishes to continue. He first presses Y (from the nature of the GET command this does not show up on the printout) and converts one more frequency. After this he must have pressed N (or any other character) since the program has terminated.

It is recommended that readers with limited experience of BASIC programming check that this simple program operates satisfactorily on their machine before progressing to those in later chapters.

Chapter 2

# Introduction to operational amplifier circuits

## 2.1 The basic amplifier

The basic amplifier may be represented by the symbol shown in Figure 2.1. The amplifier has two inputs which are denoted by $V_{i+}$ and $V_{i-}$ and a single output, $V_o$. Positive and negative power supplies of equal magnitude are normally used (although single-supply operation is possible) and are shown as $\pm V_s$ in the figure (for simplicity these connections are not normally shown on circuit diagrams). The common zero of $+V_s$ and $-V_s$ is an important reference value to which $V_{i+}$, $V_{i-}$ and $V_o$ are referred. It does not appear explicitly on the amplifier symbol since a direct connection is not required. However, one or other of the amplifier inputs may be connected to it either directly or indirectly depending on the required mode of operation.

Ideal operation of the amplifier is shown in the transfer characteristic of Figure 2.2. Here $V_i$ represents the difference between the voltages applied to the two inputs ($V_{i+}$ and $V_{i-}$). It can be seen that if $V_i$ is positive, even by only a small amount, the output $V_o$ is positive and constant, having a magnitude slightly less than that of the supply voltage (the 'output saturation voltage'). Similarly, negative values of $V_i$ produce a constant negative output.

In practice, a finite change in $V_i$ will be needed in order to change $V_o$ from one level to the other as shown by the dotted line in Figure 2.2. Also, the changeover will occur for a value of $V_i$ which is not precisely equal to zero (this effect will be discussed further in Chapter 4).

For a characteristic having a finite slope, the input/output relationship may be written as

$$V_o = A(V_{i+} - V_{i-}) \qquad (2.1)$$

where $A$ is the gain of the amplifier in the region between the two output saturation voltages. The value of $A$ is large for practical amplifiers (at least 50 000) and theoretically infinite for ideal ones. $A$ is known as the open loop gain, which is the gain of the amplifier without feedback (an external connection which makes $V_i$ depend on

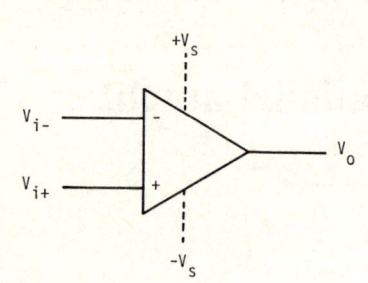

*Figure 2.1* Basic operational amplifier symbol

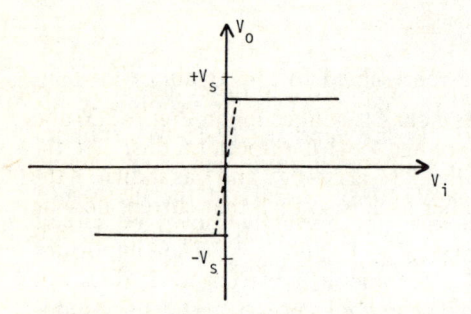

*Figure 2.2* Ideal transfer characteristic (solid line) and practical approximation (broken line) ($V_i = V_{i+} - V_{i-}$)

$V_o$ in some way). The inputs (indicated by + and − in Figure 2.1) are referred to as non-inverting and inverting respectively, for reasons which are evident from Equation (2.1).

The amplifier can be used in the basic form described above in order to distinguish between positive and negative input values. If used in this manner it would be described as a comparator, and the output levels would normally be constrained to levels suitable for connection to digital logic circuits. In the present context a continuous relationship between input and output is required and is achieved by means of feedback. Several different configurations are widely used; they are discussed in the following sections. An application of a comparator will be discussed briefly in Section 5.2. Operation without feedback is often referred to as open loop operation, which becomes closed loop operation when feedback is applied; that is, when the feedback loop is closed.

## 2.2 Inverting mode, operation as scaler and summer

The basic configuration is shown in Figure 2.3 where the resistors $R_i$ and $R_f$ are the input and feedback resistors respectively. The non-

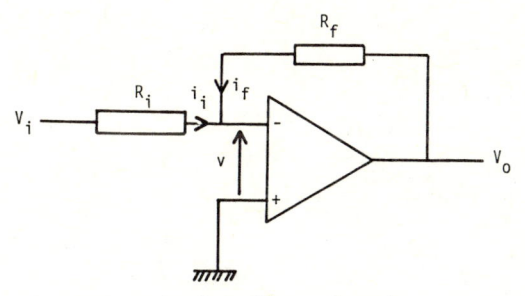

*Figure 2.3* Operational amplifier configuration

inverting input of the amplifier is connected to the common zero of the power supplies (shown as a chassis connection in Figure 2.3) and the inverting input has a voltage $v$ with respect to this. Let the currents in the input and feedback resistors be $i_i$ and $i_f$ as shown. If the input resistance of the amplifier itself is so high that current flowing into the inverting input may be neglected (an assumption which is normally justified in practice), the currents will sum to zero:

$$i_i + i_f = 0$$

Ohm's law can be applied to each resistor:

$$\frac{V_i - v}{R_i} + \frac{V_o - v}{R_f} = 0 \tag{2.2}$$

In this configuration Equation (2.1) becomes

$$V_o = -Av$$

Hence

$$\frac{V_i + V_o/A}{R_i} + \frac{V_o + V_o/A}{R_f} = 0$$

For large values of $A$, $v$ tends to zero and this reduces to

$$\frac{V_i}{R_i} + \frac{V_o}{R_f} = 0$$

or

$$V_o = -\frac{R_f}{R_i} V_i \tag{2.3}$$

This is an important and useful result since the relationship between $V_o$ and $V_i$ (a 'gain' of $-R_f/R_i$) depends only on the values of the resistors and not on the characteristics of the amplifier itself. This is

8  Introduction to operational amplifier circuits

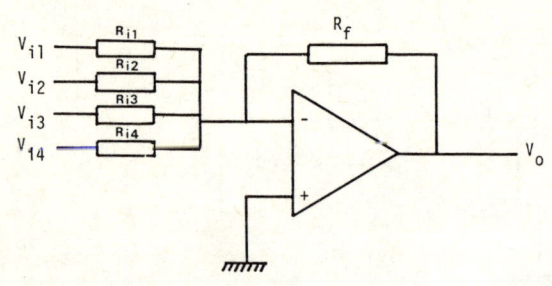

*Figure 2.4* Circuit for summing several inputs

true, of course, only when the circuit is operating under such conditions that the assumptions of very high input resistance and very high open loop gain are valid. Since $v$ has become very small, the potential of the inverting input is very close to that of the common reference. Consequently this point is often referred to as a virtual earth.

The circuit of Figure 2.3 is, therefore, capable of multiplying the input voltage by a negative constant which may be made less than, equal to, or greater than one by an appropriate choice of $R_f$ and $R_i$. This is often described as scaling. A straightforward extension to this circuit allows several input voltages to be added, and scaled if required, as shown in Figure 2.4. Summing the input and feedback currents as before yields:

$$V_o = -\left[\frac{R_f}{R_{i_1}} V_{i_1} + \frac{R_f}{R_{i_2}} V_{i_2} + \frac{R_f}{R_{i_3}} V_{i_3} + \frac{R_f}{R_{i_4}} V_{i_4}\right] \qquad (2.4)$$

Notice that a change in $R_f$ alters the scaling of all inputs and that input resistance values can be used to define the scaling of individual inputs. The number of inputs is not limited to four, of course, but a practical limit is imposed by the fact that the sum of all input currents must be balanced by the amplifier output current flowing through $R_f$. The junction of the input and feedback resistors at the non-inverting input is often referred to as the 'summing junction' (it is also the virtual earth point).

Since the basic circuit negates the input voltages, subtraction may be readily achieved by using a second operational amplifier for this purpose. For example

$$V_o = V_{i_1} - 2V_{i_2} + 3V_{i_3} - 4V_{i_4}$$

could be realized by

$$V_o = -[(2V_{i_2} + 4V_{i_4}) - (V_{i_1} + 3V_{i_3})]$$

as shown in Figure 2.5.

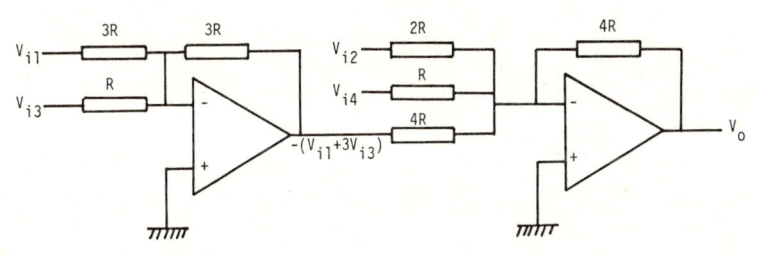

Figure 2.5 Realization of $V_{i_1} - 2V_{i_2} + 3V_{i_3} - 4V_{i_4}$ using two operational amplifiers

Since the scaling factors are defined by the ratio of two resistors an arbitrary choice of one value must be made. However, there are practical limits to the values that may be employed. For Equations (2.3) and (2.4) to hold under all conditions, the amplifier must be capable of providing an output current that exceeds $V_o/R_f$ when $V_o$ is at its maximum value. This imposes a minimum value on $R_f$; attempts to use a lower value cause the summing junction to be 'pulled' away from its value close to zero and the circuit ceases to operate correctly. At the other extreme, very high values of feedback resistance (typically in excess of a megohm) should normally be avoided since the offset error due to bias current can become large (see Chapter 4).

The output resistance of operational amplifiers is very low since an already low value (a few tens of ohms) for the basic amplifier is reduced by a factor of the order of $A$ in the operational amplifier configuration. This means that the amplifier output closely resembles an ideal voltage source (the voltage is not affected by any reasonable load that may be connected). The magnitude of this voltage depends only on the input(s) and associated passive components—a 'controlled voltage source'. Amplifiers of this kind can, therefore, be connected in cascade without interaction. This in turn means that complex functions may be realized by the interconnection of 'building blocks' that perform the required basic operations.

Note also, however, that determination of the minimum practical value of feedback resistance discussed above must also take account of the input resistors of succeeding amplifiers. This means that, when an operational amplifier drives one or more others, the minimum load resistance criterion must be applied to the value of the feedback resistor and all driven input resistances in parallel. The input resistance of an operational amplifier (at the summing junction) is very low and, therefore, very nearly equal to $R_i$ when measured at the remote end of $R_i$.

Program P(2.1) allows suitable input and feedback resistance values to be determined for a particular application. It first requests

READY.

```
10 REM PROGRAM P(2.1)
20 REM PROGRAM TO DETERMINE INPUT AND FEEDBACK
25 REM RESISTANCES FOR OP AMPS
30 REM CALCULATE MIN FEEDBACK RESISTANCE
40 PRINT ""
45 DIM R(10):DIM G(10)
50 PRINT"SUPPLY VOLTAGE":INPUT VS
60 PRINT"MAX AMPLIFIER OUTPUT CURRENT (MA)":INPUT IM
70 PRINT"LOAD TO BE DRIVEN (KOHMS)":INPUT RL
80 IF IM*RL<=VS THEN PRINT"LOAD TOO SMALL":GOTO 70
90 RM=(RL*VS)/(IM*RL-VS)
100 PRINT"MIN. RF IS "RM"KOHMS CHOOSE A VALUE"
110 INPUT RF
120 IF RF>=RM GOTO 140
130 PRINT "RF TOO SMALL":GOTO 100
140 PRINT"NUMBER OF INPUTS REQUIRED":INPUT N
150 FOR A=1TON
160 PRINT"GAIN REQURED FOR INPUT#"A:INPUT G(A)
170 R(A)=RF/G(A)
180 PRINT"INPUT RESISTANCE #"A"="R(A)"KOHMS"
190 NEXT A
200 PRINT"FEEDBACK RESISTANCE="RF"KOHMS"
210 PRINT"IF INPUT RESISTANCES TOO LOW, TYPE L"
220 GET A$:IF A$=""THEN 220
230 IF A$="L"THEN PRINT"TRY HIGHER RF":GOTO 100
```
READY.

the supply voltage ($V_s$),* the maximum output current capability of the amplifier ($I_M$) and the load resistance to be driven ($R_L$) (lines 50, 60 and 70). The total load which can be driven must exceed $V_s/I_M$ and will consist of $R_L$ in parallel with the feedback resistance. Hence the limiting case is given by:

$$\frac{R_M R_L}{R_M + R_L} = \frac{V_s}{I_M}$$

(where $R_M$ is the minimum possible feedback resistance).
Therefore

$$R_M = \frac{R_L V_s}{I_M R_L - V_s} \qquad (2.5)$$

Notice that, if a value of $R_L$ has been chosen which is beyond the capability of the amplifier, the denominator of Equation (2.5) would become negative, or zero in the limiting case. This condition is tested in line 80 of the program and an amended load resistance is requested if necessary (line 70 again). Some manufacturers do not quote a maximum output current capability in their data sheets. In this case a minimum load resistance is usually quoted; the program can readily be modified to accept this value of $R_M$ directly.

* For clarity, subscripts are used in the text, but in the program listings $V_s$, for example, will appear as VS.

If all is well, the minimum permissible feedback resistance is displayed (line 100); this will normally be an awkward value and a convenient practical one is requested. This must, of course, exceed the theoretical minimum (tested in line 120) and may be made much larger if the minimum value is particularly small.

The program then requests the required number of inputs (line 140). For ease of manipulation, the parameters associated with each input (gain and resistance) are stored in the form of arrays (R(A) and G(A)). The dimension statement (line 45) provides for a maximum of ten inputs which should be adequate. If more are required the DIM statement can be modified appropriately. Finally, the program requests the gain required for each input (line 160) and calculates the required input resistance as the ratio of the chosen feedback resistance and required gain (see Equation (2.4)) (line 170). For convenience, these operations are carried out in a FOR loop (line 150 to line 190).

If a particularly high gain has been requested, the resulting input resistance value may be too low for the particular application. Lines 210 to 230 therefore allow the process to be repeated using a higher value of feedback resistance.

A specimen run of the program is presented. Notice that the operator initially chooses a value of $R_F$ which is a convenient preferred value but not quite large enough. The smallest input resistance of 660 ohms is apparently too small and the operator has repeated the calculations by typing L (which does not show on the screen or the specimen printout as a consequence of the use of the GET command). Finally, the program is terminated by typing any character other than L.

## 2.3 Non-inverting mode, voltage follower

In some applications, the sign change associated with the inverting mode of operation is not required. A non-inverting configuration is shown in Figure 2.6(a). The potential $v$ at the inverting input may be derived from the output voltage $V_o$:

$$v = V_o \frac{R_i}{R_f + R_i}$$

(The notation $R_f$ and $R_i$ has been retained although $R_i$ is not now directly associated with the input. As before, current flowing into the amplifier is assumed to be negligible.)

From Equation (2.1), $V_o = A(V_i - v)$. Therefore

$$V_i - v = \frac{V_o}{A}$$

12   Introduction to operational amplifier circuits

## Specimen run

```
SUPPLY VOLTAGE
? 15
MAX AMPLIFIER OUTPUT CURRENT (MA)
? 10
LOAD TO BE DRIVEN (KOHMS)
? 4.7
MIN. RF IS  2.203125 KOHMS CHOOSE A VALUE
? 2.2
RF TOO SMALL
MIN. RF IS  2.203125 KOHMS CHOOSE A VALUE
? 3.3
NUMBER OF INPUTS REQUIRED
? 3
GAIN REQURED FOR INPUT# 1
? 1
INPUT RESISTANCE # 1 = 3.3 KOHMS
GAIN REQURED FOR INPUT# 2
? 2
INPUT RESISTANCE # 2 = 1.65 KOHMS
GAIN REQURED FOR INPUT# 3
? 5
INPUT RESISTANCE # 3 = .66 KOHMS
FEEDBACK RESISTANCE= 3.3 KOHMS
IF INPUT RESISTANCES TOO LOW, TYPE L
TRY HIGHER RF
MIN. RF IS  2.203125 KOHMS CHOOSE A VALUE
? 22
NUMBER OF INPUTS REQUIRED
? 3
GAIN REQURED FOR INPUT# 1
? 1
INPUT RESISTANCE # 1 = 22 KOHMS
GAIN REQURED FOR INPUT# 2
? 2
INPUT RESISTANCE # 2 = 11 KOHMS
GAIN REQURED FOR INPUT# 3
? 5
INPUT RESISTANCE # 3 = 4.4 KOHMS
FEEDBACK RESISTANCE= 22 KOHMS
IF INPUT RESISTANCES TOO LOW, TYPE L
```

*Figure 2.6(a)* Non-inverting feedback amplifier

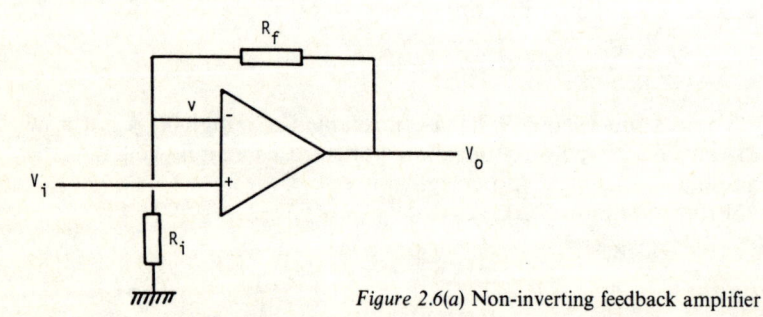

As $A$ becomes very large $V_i$ becomes nearly equal to $v$ and

$$V_i = \frac{V_o R_i}{R_f + R_i}$$

Therefore

$$V_o = V_i \left(\frac{R_f + R_i}{R_i}\right) = V_i \left(1 + \frac{R_f}{R_i}\right) \qquad (2.6)$$

As before, the relationship between input and output voltages depend only on $R_f$ and $R_i$. This time the constant of proportionality is positive and of a slightly different form (compare Equations (2.3) and (2.6)); notice that the gain cannot be less than unity. Several input resistors can be used, to provide summation without inversion (as shown in Figure 2.6(b)). However, the scaling factors are more complex; it can be shown that, for this circuit:

$$V_o = V_{i_1}\left[\frac{R_{i_2}(R_f + R_i)}{R_i(R_{i_1} + R_{i_2})}\right] + V_{i_2}\left[\frac{R_{i_1}(R_f + R_i)}{R_i(R_{i_1} + R_{i_2})}\right]$$

Since feedback action causes the potential of the inverting input to follow closely that of the non-inverting input, negligible current is drawn from the source and the circuit has a very high effective input resistance. This, combined with a very low output resistance, as for the inverting configuration, makes this circuit a particularly useful one.

In particular, if $R_f$ is made equal to zero, the gain becomes equal to unity regardless of the value of $R_i$ (Equation (2.6)) which can be omitted. The result is a simple, but highly effective, unity gain buffer amplifier usually known as a voltage follower (see Figure 2.7). It is

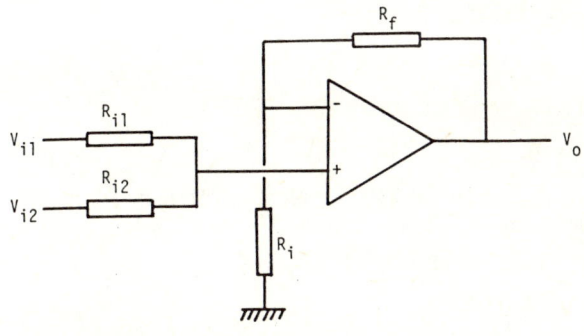

*Figure 2.6(b)* Non-inverting summer

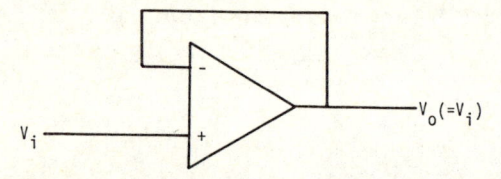

14  Introduction to operational amplifier circuits

*Figure 2.7* Unity gain buffer amplifier, or voltage follower

particularly suitable for preventing interaction between cascaded sections of a circuit.

Suitable component values (for applications other than unity gain) may be determined using program P(2.2), which is similar to P(2.1). The main differences include a test that the requested gain is greater than unity (line 150), a single input resistor since the more complex summing mode is not used at this stage (see Exercise (2.2), Section 2.7) and an appropriate equation to calculate the value of $R_i$ (line 170).

A specimen run of the program is shown and should be self-explanatory.

## 2.4 Differential mode

The inverting and non-inverting configurations, discussed in the previous sections, can be combined in order to obtain a difference signal using a single amplifier. A suitable circuit is shown in Figure 2.8.

Summing currents at the inverting input, as before:

$$\frac{V_a - v_-}{R_a} + \frac{V_o - v_-}{R_f} = 0$$

By potentiometer action:

$$v_+ = V_b R'$$

where

$$R' = R_g/(R_b + R_g)$$

(in both cases current flow into the amplifier is neglected).

The high open loop gain of the amplifier ensures that $v_+$ will be very nearly equal to $v_-$ and hence:

$$\frac{V_a - V_b R'}{R_a} + \frac{V_o - V_b R'}{R_f} = 0$$

## Differential mode

```
10 REM PROGRAM P(2.2)
20 REM PROGRAM TO DETERMINE INPUT AND FEEDBACK
25 REM RESISTANCES FOR OP AMPS
30 REM USED IN THE NON-INVERTING CONFIGURATION
40 PRINT""
50 PRINT"SUPPLY VOLTAGE":INPUT VS
60 PRINT"MAX AMPLIFIER OUTPUT CURRENT (MA)":INPUT IM
70 PRINT"LOAD TO BE DRIVEN (KOHMS)":INPUT RL
80 IF IM*RL<=VS THEN PRINT"LOAD TOO SMALL":GOTO 70
90 RM=(RL*VS)/(IM*RL-VS)
100 PRINT"MINIMUM RF+RI IS "RM" KOHMS CHOOSE RF"
110 INPUT RF
120 IF RF>=RM GOTO 140
130 PRINT "RF TOO SMALL":GOTO 100
140 PRINT"GAIN REQUIRED":INPUT G
150 IF G>1 THEN GOTO 170
160 PRINT"GAIN MUST BE > 1":GOTO140
170 RI=RF/(G-1)
180 PRINT"INPUT RESISTANCE(RI)="RI"KOHMS"
190 PRINT"FEEDBACK RESISTANCE="RF"KOHMS"
```

*Specimen run*

```
SUPPLY VOLTAGE
? 15
MAX AMPLIFIER OUTPUT CURRENT (MA)
? 7.5
LOAD TO BE DRIVEN (KOHMS)
? 4.7
MINIMUM RF+RI IS  3.48148148  KOHMS CHOOSE RF
? 3.3
RF TOO SMALL
MINIMUM RF+RI IS  3.48148148  KOHMS CHOOSE RF
? 4.7
GAIN REQUIRED
? 3
INPUT RESISTANCE(RI)= 2.35 KOHMS
FEEDBACK RESISTANCE= 4.7 KOHMS
```

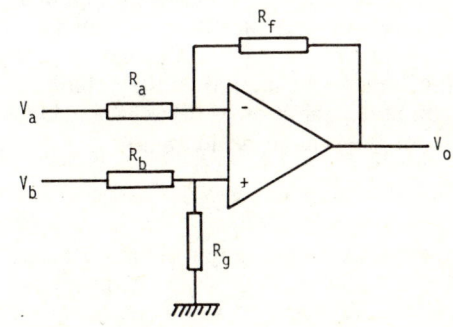

*Figure 2.8* Difference, or differential, amplifier configuration

Therefore

$$\frac{V_o}{R_f} = V_b \left[ \frac{R'}{R_a} + \frac{R'}{R_f} \right] - \frac{V_a}{R_a}$$

Substituting for $R'$ and multiplying by $R_f$:

$$V_o = V_b \left[ \frac{R_g}{(R_b + R_g)} \frac{(R_a + R_f)}{R_a} \right] - V_a \frac{R_f}{R_a} \qquad (2.7)$$

Notice that the (negative) gain with respect to input $V_a$ can be varied over a wide range by choice of $R_f$ and $R_a$ as before. However, for a fixed ratio of $R_f$ and $R_a$, the gain with respect to input $V_b$ cannot exceed $(1 + R_f R_a)$ which is obtained when $R_b = 0$ (a direct connection).

A particularly useful case arises when $R_a = R_b$ and $R_f = R_g$; Equation (2.7) then reduces to:

$$V_o = (V_b - V_a) \frac{R_f}{R_a} \qquad (2.8)$$

and the output is directly proportional to the difference of the two inputs.

Suitable values of $R_f$ and $R_a$ can be determined using program P(2.1) with only a single input. If required the program could be extended to cope with the unequal gain case of Equation (2.7). Attempts to exceed the gain specified above for input $V_b$ will result in a negative value for $R_b$ and a suitable check should be included.

The output resistance of the differential amplifier will be very low, as in the configurations discussed previously. The definition of input resistance is more complicated since two distinct modes of operation are possible:

(1) Common mode operation, where an input potential is applied to $V_a$ and $V_b$ simultaneously. Neglecting current flow into the amplifier, the approximate effective input resistance is $(R_a + R_f)$ in parallel with $(R_b + R_g)$ or $(R_a + R_f)/2$ for the symmetrical case of Equation (2.8).
(2) Differential mode operation, where the input potentials change in opposite senses; since $v_-$ is close to $v_+$ the approximate effective input resistance is $R_a + R_b$ or $2R_a$ for the symmetrical case.

## 2.5 Common mode rejection

Signals which appear simultaneously on both inputs of a differential amplifier are described as 'common mode'. Those which appear as a difference between the two inputs are described as 'differential mode' (sometimes 'direct' or 'series' mode).

## Common mode rejection

In many practical applications, such as amplification of the output from a transducer which would be connected between $V_a$ and $V_b$ of Figure 2.8, the required signal is in differential mode. Any common mode signal which occurs on the two inputs is caused by the pick-up of interference on the leads connecting the transducer (and perhaps the transducer itself). It is clearly important that the latter signal should make the smallest possible contribution to the amplifier output signal; hence the importance of common mode rejection. Equation (2.8) implies that the common mode gain of the amplifier should be precisely zero and the differential mode gain should be $R_f/R_a$ as required.

Unfortunately, zero common mode gain assumes perfect matching of $R_a$ to $R_b$ and $R_f$ to $R_g$. In practice this will not be so and other spurious effects within the amplifier itself will combine to make the common mode gain small but *not* zero. Clearly, the smallness of this gain is a measure of the merit of the amplifier. This parameter is usually specified in inverse form as a 'common mode rejection ratio' (CMRR) with respect to the differential mode gain and expressed in decibels, hence

$$\text{CMRR} = 20 \log_{10} \frac{\text{(Differential mode gain)}}{\text{(Common mode gain)}} \qquad (2.9)$$

The common mode rejection ratio is normally specified in manufacturers' data sheets for open loop operation and for one or more values of closed loop operation. The quoted value will be further degraded in many applications by component mismatch. For this reason $R_g$, for example, in Figure 2.8 may consist of a fixed and a small variable resistance in series so that the latter can be adjusted for optimum common mode rejection.

In addition, although the common mode gain may be very small, the permissible magnitude of such a signal which may be applied to the amplifier inputs will have practical limits as specified by the manufacturer.

In general, therefore, the amplifier output will contain components due to both differential and common mode signals and can be written as

$$V_o = A_{\text{DM}}(V_b - V_a) + \frac{A_{\text{CM}}}{2}(V_b + V_a) \qquad (2.10)$$

where the subscripts DM and CM refer to differential and common mode gains respectively. The factor $\frac{1}{2}$ is normally included so that, for $V_b = V_a$, the common mode gain becomes $A_{\text{CM}}$ (rather than $2A_{\text{CM}}$).

The values of $A_{\text{DM}}$ and $A_{\text{CM}}$ could be measured experimentally or derived for a particular circuit taking due account of all relevant component tolerances. However, both these approaches yield gains with respect to $V_a$ and $V_b$; that is, an expression of the form:

$$V_o = A_b V_b - A_a V_a \tag{2.11}$$

(compare this with Equation (2.7)).

In order to determine the CMRR it is necessary to obtain $A_{DM}$ and $A_{CM}$ from $A_a$ and $A_b$ as follows. From Equation (2.10):

$$V_o = V_a \left( \frac{A_{CM}}{2} - A_{DM} \right) + V_b \left( \frac{A_{CM}}{2} + A_{DM} \right)$$

From Equation (2.11):

$$A_a = -\left( \frac{A_{CM}}{2} - A_{DM} \right)$$

and

$$A_b = \left( \frac{A_{CM}}{2} + A_{DM} \right)$$

Adding:

$$(A_a + A_b) = 2A_{DM}$$

Subtracting:

$$(A_b - A_a) = A_{CM}$$

Hence, from Equation (2.9), the common mode rejection ratio is given by

$$\text{CMRR} = 20 \log_{10} \frac{(A_b + A_a)}{2(A_b - A_a)} \tag{2.12}$$

This is realized by program P(2.3) which should be largely self-explanatory. In the form presented, the program accepts gains as pure ratios and determines the CMRR in decibels.

If the *a* and *b* channel gains are equal, a divide-by-zero error will occur when Equation (2.12) is evaluated. This case is tested for in line 80 and indicated by jumping to line 160. Line 90 ensures that the difference in gains is positive in order to avoid attempting to evaluate a negative logarithm. If the computer used evaluates natural logarithms, as is often the case, a division by $\log_e 10$ is required to convert to base 10 (line 110).

Three specimen runs of the program are presented; the final one illustrates the very close matching of the *a* and *b* channel gains which is required in order to obtain a really high common mode rejection.

```
10 REM PROGRAM P(2.3)
20 REM PROGRAM TO DETERMINE CMRR
30 PRINT"A CHANNEL GAIN"
40 INPUT AA
50 PRINT"B CHANNEL GAIN"
60 INPUT AB
70 D=(AB-AA)
80 IF D=0 THEN 160
90 IF D<0THEN D=-D
100 R1=(AA+AB)/(2*D)
110 R2=20*(LOG(R1)/LOG(10))
120 PRINT"CMRR="R2"DB"
130 PRINT"PRESS ANY KEY TO REPEAT"
140 GET A$:IF A$=""THEN140
150 GOTO 20
160 PRINT "GAINS EQUAL CMRR IS INFINITE"
170 GOTO 130
```

*Specimen run*

```
A CHANNEL GAIN
? 100
B CHANNEL GAIN
? 99
CMRR= 39.9564616 DB
PRESS ANY KEY TO REPEAT
A CHANNEL GAIN
? 50
B CHANNEL GAIN
? 50
GAINS EQUAL CMRR IS INFINITE
PRESS ANY KEY TO REPEAT
A CHANNEL GAIN
? 1000
B CHANNEL GAIN
? 999.99
CMRR= 100.000172 DB
PRESS ANY KEY TO REPEAT
```

## 2.6 Instrumentation amplifier

Although useful, the differential amplifier discussed in the previous section is unsuitable for applications involving high impedance sources because of its relatively low input resistance (determined essentially by the input components themselves).

The obvious remedy is to insert voltage followers (Figure 2.7) in each input path. Having incurred the expense of two extra amplifiers, it is tempting to operate these at a gain greater than unity (Figure 2.6(a)). Unfortunately, in order to preserve common mode rejection, this approach would involve the matching of no less than four pairs of resistors. A clever interconnection of the buffer amplifiers avoids this requirement. The resulting circuit is shown in Figure 2.9 and is

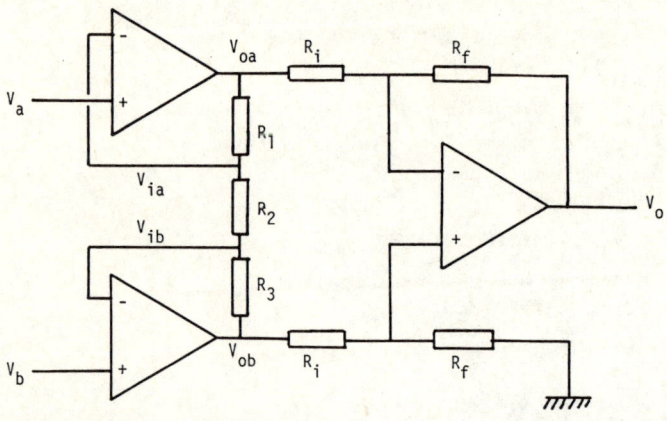

20  Introduction to operational amplifier circuits

*Figure 2.9* Instrumentation amplifier

known as an instrumentation amplifier because of its widespread use in measurement systems.

If the input current to the amplifiers can be neglected, the same current will flow through $R_1$, $R_2$ and $R_3$; let this be $I$. With voltages as indicated in the figure:

$$I = \frac{V_{oa} - V_{ia}}{R_1} = \frac{V_{ia} - V_{ib}}{R_2} = \frac{V_{ib} - V_{ob}}{R_3}$$

For the first pair:

$$\frac{V_{oa}}{R_1} = V_{ia}\left(\frac{1}{R_2} + \frac{1}{R_1}\right) - \frac{V_{ib}}{R_2}$$

Similarly, for the second pair:

$$\frac{V_{ob}}{R_3} = V_{ib}\left(\frac{1}{R_3} + \frac{1}{R_2}\right) - \frac{V_{ia}}{R_2}$$

In view of the high gain of the buffer amplifier

$$V_{ia} \simeq V_a \quad \text{and} \quad V_{ib} \simeq V_b$$

Hence:

$$V_{oa} - V_{ob} = V_a\left(1 + \frac{R_1}{R_2} + \frac{R_3}{R_2}\right) - V_b\left(1 + \frac{R_1}{R_2} + \frac{R_3}{R_2}\right)$$

Notice that the coefficients are the same and the expression reduces to

$$V_{oa} - V_{ob} = (V_a - V_b)\left(1 + \frac{R_1 + R_3}{R_2}\right) \tag{2.13}$$

which has zero common mode gain without any requirement for component matching. $R_2$ can conveniently be made variable in order to control the gain.

The overall gain will be given by

$$V_o = (V_a - V_b)\left(1 + \frac{R_1 + R_3}{R_2}\right)\frac{R_f}{R_i}$$

and the requirement for component matching in the output stage remains; as before, one of these can be made adjustable in order to optimize common mode rejection.

Instrumentation amplifiers are available in the form of integrated circuit modules; one well-established device (Analog Devices AD 524) has built-in resistances for the equivalent of $R_2$. These can be selected by means of a link to give gains of 10, 100 or 1000; intermediate values can be obtained by using a resistor in place of the link.

If $R_2$ is removed, the input stages become unity gain buffers. However, if $R_2$ is short-circuited, Equation (2.13) suggests that the gain of the input stage should become infinite. This will not be the case, of course, since, for values of closed loop gain which approach the open loop value, the assumptions $V_{ia} \simeq V_a$ and $V_{ib} \simeq V_b$ are no longer valid. In practice, the circuit tends to oscillate when attempts are made to obtain very high gains.

Program P(2.4) calculates the required value of $R_2$ using Equation (2.13), for a specified value of buffer stage gain, given nominal (and equal) values of $R_1$ and $R_3$. Notice that the required gain G must be greater than unity. Tests are applied at lines 90 and 100 with appropriate comments at lines 160 and 180.

```
10 REM PROGRAM P(2.4)
20 REM PROGRAM TO DETERMINE RESISTANCE VALUE
30 REM FOR INSTRUMENTATION AMPLIFIER
40 PRINT"⌂"
50 PRINT"REQUIRED VALUE OF R1 AND R3 (KOHMS)"
60 INPUT R1
70 PRINT"REQUIRED GAIN"
80 INPUT G
90 IF G=1THEN160
100 IF G<1THEN 180
110 R2=(2*R1)/(G-1)
120 PRINT"REQUIRED VALUE OF R2 IS"R2"KOHMS"
130 PRINT"PRESS ANY KEY TO REPEAT"
140 GET A$:IF A$=""THEN 140
150 GOTO 50
160 PRINT"R2 NOT NEEDED FOR UNITY GAIN CASE"
170 GOTO 130
180 PRINT "CANNOT REALISE GAINS LESS THAN UNITY"
190 GOTO 130
```

## Specimen run

```
REQUIRED VALUE OF R1 AND R3 (KOHMS)
? 4.7
REQUIRED GAIN
? 15
REQUIRED VALUE OF R2 IS .671428571 KOHMS
PRESS ANY KEY TO REPEAT
REQUIRED VALUE OF R1 AND R3 (KOHMS)
? 4.7
REQUIRED GAIN
? .9
CANNOT REALISE GAINS LESS THAN UNITY
PRESS ANY KEY TO REPEAT
REQUIRED VALUE OF R1 AND R3 (KOHMS)
? 100
REQUIRED GAIN
? 15
REQUIRED VALUE OF R2 IS 14.2857143 KOHMS
PRESS ANY KEY TO REPEAT
```

Three specimen runs of the program are presented and should be self-explanatory. The program relates only to the buffer stages; if required, it could be extended to include the gain provided by the differential output stage. In this case an overall gain of less than unity would be possible but is unlikely to be required in practice.

## 2.7 Exercises

**(2.1)** Modify program P(2.1) so that, when L is typed in order to permit the use of a higher feedback resistance, the required number of inputs and associated gains are re-used without the need to re-enter them.

**(2.2)** Derive the input/output relationship for the non-inverting summer of Figure 2.6(b). Hence, extend program P(2.2) to include non-inverting summers.

**(2.3)** A differential amplifier, of the form shown in Figure 2.8, has input and feedback resistors of $1\,k\Omega$ and $10\,k\Omega$ respectively. If the resistors used have a tolerance of $\pm 2$ per cent, deduce the $a$ and $b$ channel gains in the worst case condition (with respect to common mode rejection). Hence, use program P(2.3) to determine the common mode rejection ratio. (Assume that the amplifier itself does not further degrade common mode rejection.)

**(2.4)** Extend program P(2.3) to accept gains as either a ratio or a decibel value.

**(2.5)** Using Equations (2.1) and (2.2) determine the input/output relationship for an operational amplifier when the open loop gain A is

finite. Hence, extend program P(2.1) to determine:
(i) the error in closed loop gain caused by the finite open loop gain,
(ii) the actual value of feedback resistance required to obtain the specified closed loop gain with a finite open loop gain.

(**2.6**) Extend program P(2.4) by incorporating the routine, given in Appendix A, to determine the nearest available preferred value for $R_2$. Hence, determine the error involved in using this value.

Chapter 3
# Frequency response

## 3.1 Open loop behaviour, compensation

The circuits discussed in the previous chapter all depended on the assumption that the open loop gain $A$ remains very large (ideally infinite) under all operating conditions. In practice this cannot be true for all frequencies. For stable operation with the feedback configurations used the high gain *must* be preserved for low frequencies, including D.C. (zero frequency).

However, there must be an upper frequency limit above which the gain decreases with frequency. Although inevitable, this is not really a disadvantage since amplification of signals outside the range of frequencies required merely increases the noise content.

The reduction in gain at high frequencies is caused essentially by stray capacitances associated with the amplifier. The simplest way of modelling this effect is a single low pass filter as shown in Figure 3.1. This first order model is adequate for many applications although a more accurate representation would include additional filter elements giving a high order transfer function.

The capacitance $C$ of Figure 3.1 represents the total effect of all stray capacitances associated with the amplifier together with additional capacitance which may be included deliberately in order to adjust the frequency at which the gain begins to fall off. This additional capacitance may be internal to the integrated circuit amplifier (in which case it would be realized as the barrier capacitance of a reverse-biased diode) or external, in which case it would be an orthodox capacitor. Both approaches are useful; the former is known as internal compensation and the latter as external compensation.

Internal compensation has the advantage that stability is guaranteed under all operating conditions and an external capacitor is not required. The disadvantage is that the available open loop bandwidth has been determined by the device manufacturer and cannot be readily changed by the user. The widely used 741 amplifier is of this type.

External compensation gives greater flexibility, but care is required

Open loop behaviour, compensation 25

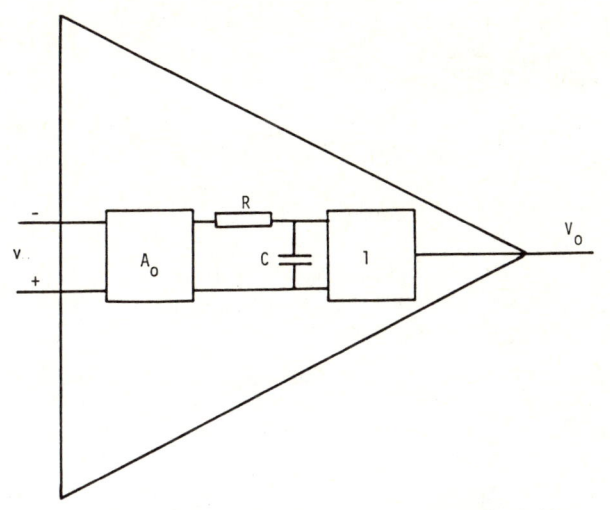

*Figure 3.1* First order model of amplifier behaviour at high frequency

since an unsuitable choice of compensating components can cause the amplifier to become unstable. The required value(s) of compensating component(s) are normally determined from manufacturers' data. The 748, for example, is essentially similar to the 741 except that it is designed for external compensation.

In Figure 3.1, $A_o$ represents the low frequency, or D.C., gain which, as usual, will be very large. The second square is a buffer whose gain may be assumed to be unity at all frequencies. Hence:

$$V_o = A_o \cdot v \cdot \frac{1/j\omega C}{R + 1/j\omega C} = A_o \cdot v \cdot \frac{1}{1 + j\omega CR} \tag{3.1}$$

where $\omega$ is the angular frequency ($=2\pi f$). For very low values of $\omega$ it can be seen that the gain tends to $A_o$, and for high values of $\omega$ it becomes small. The presence of a complex denominator means that there will also be a phase change with frequency. From Equation (3.1), the effective open loop gain is

$$A = \frac{V_o}{v} = A_o \frac{1}{1 + j\omega CR}$$

This is often written

$$A_o \frac{1}{1 + j(\omega/\omega_o)} \tag{3.2}$$

where $\omega_o = 1/CR$.

It is customary to express the gain in decibels, hence:

Gain (dB) = $20 \log_{10} |A|$

$$= 20 \log_{10} \frac{A_o}{\sqrt{[1 + (\omega/\omega_o)^2]}} \qquad (3.3)$$

It is useful to consider three cases:

(1) $\omega \ll \omega_o$. The gain tends to $20 \log_{10} A_o$ which is the D.C. gain. So the graph of gain versus frequency will be a horizontal straight line under this condition.

(2) $\omega \gg \omega_o$. This implies that $\omega/\omega_o$ is much larger than unity so (3.3) becomes

$$\text{Gain (dB)} = 20 \log_{10} \frac{A_o}{(\omega/\omega_o)} \qquad (3.4)$$

which may be written:

$$20 \log_{10} A_o + 20 \log_{10} \omega_o - 20 \log \omega \qquad (3.5)$$

The first two terms are constant with respect to $\omega$ and will therefore disappear if the ratio of the gains at two different frequencies is taken (since a ratio implies the subtraction of logarithms). Consider two frequencies $\omega_1 = n\omega_o$ and $\omega_2 = 10n\omega_o$, where $n$ is large, so that both $\omega_1$ and $\omega_2 \gg \omega_o$. From Equation (3.5), the ratio of the gains at these two frequencies is given by

Gain (dB) at $\omega_2$ − Gain (dB) at $\omega_1$

$= -20 \log_{10} 10n\omega_o - (-20 \log_{10} n\omega_o)$

$= -20 \log_{10} 10 = -20$ dB

The gain, therefore, decreases by 20 dB for each factor of ten increase in frequency. This is usually described as a slope, or 'roll-off', of 20 dB per decade. A similar argument, based on a doubling of the frequency, gives a slope of 6 dB per octave (where $\log_{10} 2$ has been approximated to 0.3). So, at high frequencies, provided a logarithmic scale is used for both gain (by the use of dB) and frequency, a straight line relationship is obtained with a (negative) slope as determined above. From Equation (3.5) this line will intersect the constant, low frequency, gain at $\omega = \omega_o = 2\pi f_o$ as shown in Figure 3.2.

(3) $\omega = \omega_o$. Clearly, for this condition neither of the two previous inequalities is valid.

Open loop behaviour, compensation 27

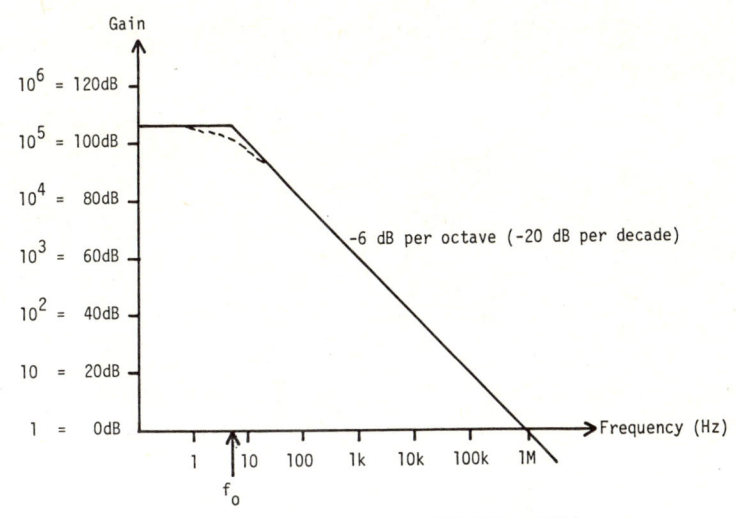

*Figure 3.2* Bode plot for the open loop 741 amplifier ($f_0 = 5\,\text{Hz}$)

From Equation (3.3):

$$\text{Gain (dB)} = 20 \log_{10} \frac{A_0}{\sqrt{2}} \tag{3.6}$$

$$= 20 \log_{10} A_0 - 20 \log_{10} \sqrt{2}$$

$$= \text{D.C. gain} - 3\,\text{dB} \tag{3.7}$$

So the straight line approximations of Figure 3.2, usually known as the 'Bode plot', are in error by 3 dB at $\omega = \omega_o$ (i.e. $f = f_o$) and the actual characteristic follows the dotted line in the figure. This error is sufficiently small to be neglected in many applications; in any case, the straight lines can be drawn very simply once $\omega_o$ is known and the required correction at the intersection point can be readily incorporated.

The frequency $\omega_o$ and its equivalent in hertz ($f_o = \omega_o/2\pi$) is known variously as the 'turnover frequency', the 'break frequency', the '3 dB point' (from Equation (3.7)) and the 'half power point' (since Equation (3.6) relates to voltage gain and $1/\sqrt{2}$ becomes squared when calculating power gain).

Figure 3.2 clearly illustrates the important concept of a constant 'Gain-bandwidth product' (GB). For all frequencies above $f_o$ the product of gain and frequency is constant; for example:

$$10^5 \times 10\,\text{Hz} = 10^3 \times 1\,\text{kHz} = 1 \times 1\,\text{MHz} = 1\,\text{MHz}$$

28    Frequency response

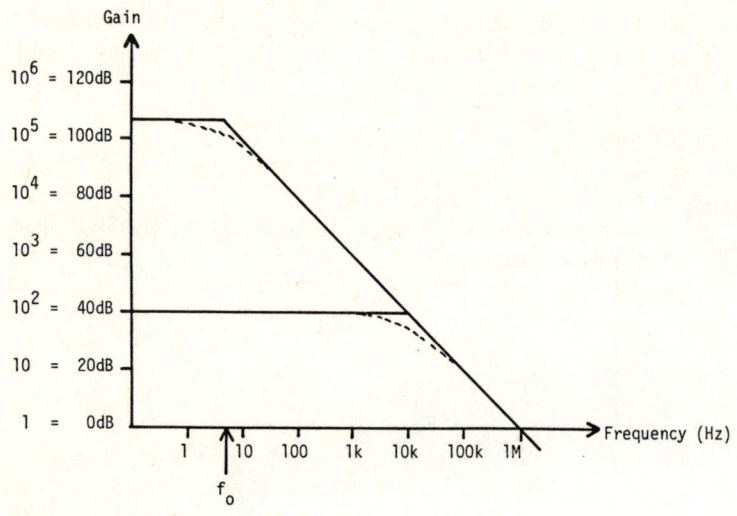

*Figure 3.3* Closed loop gain of 100 (40 dB) superimposed on the open loop characteristic

## 3.2 Closed loop response, rise time

The ideas developed in the previous section relate to the amplifier itself. What happens when feedback is applied, as in Figure 2.3, in order to obtain a defined closed loop gain?

Clearly, at very low frequencies the assumption that $A$ is very large will still be valid. But, as the frequency increases, this will become progressively less true until the required closed loop gain becomes equal to and then exceeds the available open loop gain. This is shown in Figure 3.3 where a required closed loop gain of 100 has been superimposed on the characteristic of Figure 3.2.

At approximately 10 kHz in this case the ideal closed loop gain characteristic intersects the open loop one; above this frequency the closed loop gain will clearly fall off since it cannot exceed the open loop value. In fact, a detailed analysis shown that the actual closed loop gain is 3 dB less than that predicted by the intersection of the two characteristics, as shown by the dotted curve in Figure 3.3.

Consideration of Figure 3.3 shows that there is a trade-off between available closed loop gain and required bandwidth. If bandwidth is defined by a 3 dB drop in the closed loop value it can be seen (from Figure 3.3) that:

$$\text{(Closed loop gain)} \times \text{(Closed loop bandwidth)} = GB \quad (3.8)$$

Large signal operation, slew rate and full power bandwidth     29

where GB is the gain-bandwidth product as defined in the previous section. For a given bandwidth, therefore, the maximum available closed loop gain follows immediately from the gain-bandwidth product of the amplifier to be used. For example, if a 741 (GB = 1 MHz) is to be used in an audio frequency application where a bandwidth of 20 kHz is needed, Equation (3.8) shows that the maximum available closed loop gain is 50.

It should be noted that the bandwidth as defined by the 3 dB point (Equation (3.7)) actually represents an error of $1/\sqrt{2}$ (Equation (3.6)), or approximately 30 per cent at the maximum frequency. Clearly, for precise applications such as instrumentation, much less bandwidth will be available. Acceptable gain errors for such applications are typically 1 or even 0.1 per cent. The available bandwidth in these cases is readily determined from the 20 dB per decade roll-off of the amplifier.

For 1 per cent error the open loop gain must be at least 100 times the required closed loop gain. This corresponds to 40 dB or two decades so the available bandwidth is one-hundredth of the 3 dB value. Similarly, 0.1 per cent error reduces the bandwidth to one-thousandth. Applying these values to the previous example of a 741 with a closed loop gain of 50, for a 1 per cent gain accuracy the bandwidth becomes 200 Hz and for 0.1 per cent, 20 Hz. This is a drastic reduction compared with the apparent implications of a 1 MHz gain-bandwidth product.

Although bandwidth is a convenient and widely used means of specifying an amplifier's behaviour with respect to frequency, the concept of rise time is appropriate in circuits designed to handle square pulses. Since the precise start and finish of the rising edge are not well defined, it is customary to specify the rise time as the time between 10 and 90 per cent of the steady state output in response to an ideal step input, as shown in Figure 3.4.

The low pass filter effect of the amplifier implies that for narrow bandwidth amplifiers the rise time will be relatively long, and vice versa. The relationship can be calculated analytically but in practice the empirical relationship

$$\text{Rise time} = \frac{0.35}{\text{Gain-bandwidth product}} \qquad (3.9)$$

(where rise time is defined as in Figure 3.4) is found to be particularly convenient.

## 3.3 Large signal operation, slew rate and full power bandwidth

The discussions on open and closed loop bandwidth and rise time in

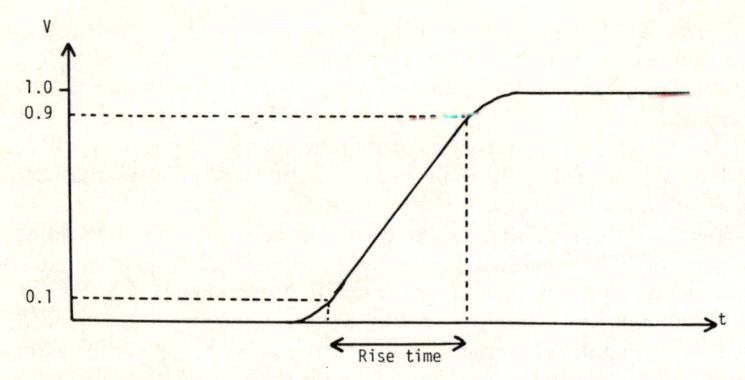

*Figure 3.4* Definition of rise time

the previous section related to 'small signal operation'. Although not specifically defined, this is generally assumed to refer to output signals of less than about 1 volt peak to peak.

For large signal operation, the output signal magnitude can approach the limits imposed by the power supply; for example, a swing of $\pm 12$ volts could be obtained with a conventional $\pm 15$ volt power supply. Under these conditions the factors that determine the effective bandwidth and rise time are somewhat different.

The simple model of Figure 3.1 is still appropriate. However, high frequency operation is now limited by the ability of amplifier $A_o$ to provide sufficient current to charge capacitor $C$ at the required rate. The corresponding rate of change of voltage at the output ($dV_o/dt$) when $C$ is being charged at its maximum rate is called the 'slew rate', and an amplifier operating in this mode is said to be 'slew rate limited'.

For example, the familiar 741 amplifier has an effective value of $C$ of 30 pF, and 15 $\mu$A (determined by the internal configuration of the amplifier) is available to charge this.

Now $Q = CV$, where $V$ is the instantaneous voltage across any capacitor; by differentiating:

$$\frac{dQ}{dt} = I = C\frac{dV}{dt}$$

Inserting the values above, $dV/dt = 0.5$ volt/$\mu$s, which is the maximum possible value and hence the slew rate.

The rise time for a 741 amplifier, for small signal operation, may be obtained from Equation (3.9) as $0.35\,\mu$s. For slew rate limited operation and an output step of height $H$, the effective rise time will be given by:

$$\text{Rise time}_{(\text{effective})} = \frac{0.8H}{\text{Slew rate}} \quad (3.10)$$

where the 0.8 arises from the 10 to 90 per cent definition of rise time (Figure 3.4). For the 741 and a 10 volt step, the effective rise time becomes 16 $\mu$s which is substantially greater than the small signal value.

Slew rate limiting will dominate when the step magnitude exceeds that value which makes the values given by Equations (3.9) and (3.10) equal. Clearly, therefore, the distinction between small and large signal operation depends on the parameters of the particular amplifier and the 1 volt value mentioned earlier is arbitrary. For the 741 amplifier the values given by Equations (3.9) and (3.10) become equal for

$$\frac{0.35}{1.0} = \frac{0.8 \times H}{0.5} \text{ volts}/\mu s$$

or $H = 0.22$ volts; the output will be slew rate limited above this level.

Slew rate limiting also has important implications in the amplification of sinusoidal signals. In particular, at relatively high frequencies and amplitudes, the maximum rate of change of the sinusoid will be limited by the slew rate of the amplifier. Bandwidth defined in this way is called the full power bandwidth and can be less than the equivalent small signal bandwidth.

Consider an output sine wave defined by:

$$V_o = E_{pk} \sin \omega t$$

where $E_{pk}$ is the peak amplitude. The rate of change is given by

$$\frac{dV_o}{dt} = \omega E_{pk} \cos \omega t$$

which has a maximum value of $E_{pk}$ when $\cos \omega t = 1$ (where the original sinusoid crosses the zero level).

The highest possible output frequency without distortion due to slew rate limiting is, therefore, given by

$$\text{Slew rate} = \omega_{max} E_{pk} = 2\pi f_{max} E_{pk}$$

Therefore

$$f_{max} = \frac{\text{Slew rate}}{2\pi E_{pk}} \quad (3.11)$$

For the 741 and a 20 volt peak to peak (10 volt peak) sine wave, the full power bandwidth is given by

$$f_{max} = \frac{0.5}{2\pi \times 10} = 7.96 \text{ kHz}$$

Notice that this value depends on the output signal magnitude and not the gain (as in the case of small signal operation). The dominant factor in determining the bandwidth depends on the required operating conditions; both small signal and full power bandwidths should always be checked for a proposed application.

Some manufacturers quote slew rate in their data sheets, some quote full power bandwidth and some quote both. Program P(3.1) provides convenient conversion between the two, taking account of the required signal amplitude; notice that conversion in either direction is possible, the choice being made in lines 50 to 80. A choice other than 'B' or 'S' causes a jump from line 90 back to line 60.

The first specimen run shows the slew rate required to maintain a peak output of 10 volts over the full audio bandwidth of 20 kHz; the second confirms the calculation above relating to the full power bandwidth of the 741 amplifier.

### 3.4 Choice of amplifier to meet a given specification

For a particular application the required closed loop gain and bandwidth will normally be known. Since the price of an operational

```
10 REM PROGRAM P(3.1)
20 REM PROGRAM TO CONVERT FULL POWER
30 REM BANDWIDTH (BW) TO SLEW RATE (SR)
40 REM AND VICE VERSA
50 PRINT "BW TO SR (B) OR SR TO BW (S)?"
60 GET A$:IF A$ = "" THEN 60
70 IF A$ = "B" THEN 1000
80 IF A$ = "S" THEN 2000
90 GO TO 60
1000 PRINT "FULL POWER BANDWIDTH (KHZ)?
1010 INPUT BW
1020 PRINT "PEAK SIGNAL VOLTAGE (VOLTS)?
1030 INPUT EP
1040 SR = 2*π*EP*BW*1E-3
1050 PRINT "SLEW RATE ="SR"VOLTS PER MICROSECOND
1060 END
2000 PRINT "SLEW RATE (VOLTS PER MICROSECOND)?"
2010 INPUT SR
2020 PRINT "PEAK SIGNAL VOLTAGE (VOLTS)?"
2030 INPUT EP
2040 BW =SR*1E3/(2*π*EP)
2050 PRINT "FULL POWER BANDWIDTH ="BW"KHZ"
2060 END
```

## Specimen run

```
BW TO SR (B) OR SR TO BW (S)?
B
FULL POWER BANDWIDTH (KHZ)?
? 20
PEAK SIGNAL VOLTAGE (VOLTS)?
? 12
SLEW RATE = 1.50796447 VOLTS PER MICROSECOND


BW TO SR (B) OR SR TO BW (S)?
S
SLEW RATE (VOLTS PER MICROSECOND)?
? .5
PEAK SIGNAL VOLTAGE (VOLTS)?
? 10
FULL POWER BANDWIDTH = 7.95774716 KHZ
```

amplifier increases markedly with available gain-bandwidth product, it is clearly desirable to use the lowest specification device (or devices) which will meet the system requirements. Program P(3.2) enables a selection to be made on this basis.

It contains a 'catalogue' of available devices (D$(J)) and their associated gain-bandwidth products (GB(I)) (lines 70 to 120). These would normally be the devices stocked by, or available to, a particular organization.

Notice that the devices included in the programs were chosen randomly (by way of illustration) and are *not* intended to represent an optimum selection in any way.

Details of the available devices and their associated GB values are stored in arrays. In some computers it may be necessary to declare the size of the arrays by means of DIM statements prior to line 70. The specimen runs were produced on a machine which assumed a default size of ten for undeclared arrays. Notice that unused elements appear with a GB of 0, which does not affect operation of the program and leaves space for additional devices.

The available gain-bandwidth products must be arranged in numerical order. This could be achieved manually during program entry but later addition, deletion or substitution of devices would be tedious; so the ordering is achieved automatically by means of a 'bubble sort' routine (lines 200 to 430).

This routine (see for example Reference 3.1) operates by examining consecutive pairs of data in the array (line 240). If any pair shows the wrong order it is reversed (lines 300 to 320). Notice that in order to achieve this, one member of the pair must be transferred to temporary storage locations (T and T$) while the other pair is swapped (line 300). Remember also that the device type (D$) must be

## 34 Frequency response

```
10 REM PROGRAM P(3.2)
20 REM PROGRAM TO FIND THE BEST AMPLIFIER
30 REM FOR A SPECIFIED GAIN AND BANDWIDTH
40 REM SET UP ARRAYS FOR GAIN BANDWDTH PRODUCT(GB) IN MHZ
60 REM OF AVAILABLE DEVICES(D$)
70 D$(1)="741":GB(1)=1
80 D$(2)="LM833":GB(2)=9
90 D$(3)="NE5534A":GB(3)=10
100 D$(4)="NE5539":GB(4)=1200
110 D$(5)="709C":GB(5)=5
120 D$(6)="4136":GB(6)=3
200 REM SORT ARRAYS INTO GB ORDER
210 REM USING A BUBBLE SORT
220 FOR I=2TO10
240 IF GB(I)>GB(I-1) THEN 300
250 NEXT I
260 IF C=0 THEN 400
270 C=0
280 GOTO 220
300 T=GB(I):T$=D$(I)
310 GB(I)=GB(I-1):D$(I)=D$(I-1)
320 GB(I-1)=T:D$(I-1)=T$
330 C=C+1
340 GOTO 250
400 PRINT "RE-ORDERED ARRAYS ARE"
410 FOR I = 1 TO 10
420 PRINT I,D$(I),"GB="GB(I)
430 NEXT I
500 REM OBTAIN REQUIRED SPECIFICATION
510 PRINT "REQUIRED GAIN"
520 INPUT G
530 PRINT "REQUIRED BANDWIDTH (KHZ)"
540 INPUT BW
550 REM CALCULATE REQUIRED GB PRODUCT (MHZ)
560 GB=(G*BW)/1000
580 I=10
590 IF GB<=GB(I) THEN 1000
600 I=I-1
605 IF I<>0 GOTO 590
610 PRINT "HIGHEST AVAILABLE GAIN BANDWIDTH PRODUCT"
620 PRINT "CANNOT MEET REQUIRED SPECIFICATION"
625 PRINT "REQUIRED GAIN BANDWIDTH PRODUCT ="GB
626 PRINT "HIGHEST AVAILABLE = "GB(1)
630 PRINT "USE AN ADDITIONAL AMPLIFIER"
640 PRINT "OR A HIGHER GAIN DEVICE"
650 STOP
1000 PRINT"LOWEST COST SUITABLE DEVICE IS A "D$(I)
1010 PRINT "WITH A GAIN BANDWIDTH PRODUCT OF"GB(I)
1020 PRINT "REQUIRED GAIN BANDWIDTH PRODUCT = "GB
1030 END
```

swapped, as well as its associated gain-bandwidth product (GB) (lines 300 to 320). This routine must be repeated (line 340) until no further swaps are required. This condition is detected by incrementing C (line 330) each time a swap is made; the routine is

## Specimen runs

```
1.  RE-ORDERED ARRAYS ARE
        1           NE5539           GB= 1200
        2           NE5534A          GB=   10
        3           LM833            GB=    9
        4           709C             GB=    5
        5           4136             GB=    3
        6           741              GB=    1
        7                            GB=    0
        8                            GB=    0
        9                            GB=    0
       10                            GB=    0
    REQUIRED GAIN
    ? 10
    REQUIRED BANDWIDTH (KHZ)
    ? 10
    LOWEST COST SUITABLE DEVICE IS A 741
    WITH A GAIN BANDWIDTH PRODUCT OF 1
    REQUIRED GAIN BANDWIDTH PRODUCT =  .1

2.  REQUIRED GAIN
    ? 100
    REQUIRED BANDWIDTH (KHZ)
    ? 10
    LOWEST COST SUITABLE DEVICE IS A 741
    WITH A GAIN BANDWIDTH PRODUCT OF 1
    REQUIRED GAIN BANDWIDTH PRODUCT =  1

3.  REQUIRED GAIN
    ? 200
    REQUIRED BANDWIDTH (KHZ)
    ? 20
    LOWEST COST SUITABLE DEVICE IS A 709C
    WITH A GAIN BANDWIDTH PRODUCT OF 5
    REQUIRED GAIN BANDWIDTH PRODUCT =  4

4.  REQUIRED GAIN
    ? 1200
    REQUIRED BANDWIDTH (KHZ)
    ? 1000
    LOWEST COST SUITABLE DEVICE IS A NE5539
    WITH A GAIN BANDWIDTH PRODUCT OF 1200
    REQUIRED GAIN BANDWIDTH PRODUCT =  1200

5.  REQUIRED GAIN
    ? 1400
    REQUIRED BANDWIDTH (KHZ)
    ? 1000
    HIGHEST AVAILABLE GAIN BANDWIDTH PRODUCT
    CANNOT MEET REQUIRED SPECIFICATION
    REQUIRED GAIN BANDWIDTH PRODUCT = 1400
    HIGHEST AVAILABLE =  1200
    USE AN ADDITIONAL AMPLIFIER
    OR A HIGHER GAIN DEVICE
```

finished when C remains at zero (line 260). If two or more devices are included that have the same GB value they will be grouped together. The program could be extended to detect this condition if required, see Exercise (3.7) (Section 3.5).

The re-ordered array is printed out by lines 400 to 430 (see the first specimen run). This is useful for test purposes but may be suppressed to save time and space (as in specimen runs 2 to 5).

The required gain and bandwidth (the 3 dB value is assumed) are obtained in lines 500 to 540 and the required gain-bandwidth product is calculated in line 560. This is then compared with the array in order to find the lowest suitable GB (lines 580 to 605) and the details printed as in specimen runs 1 to 4. Run 5 illustrates the case where the required specification cannot be met.

It is important to realize that this program is based solely on gain-bandwidth product considerations. When using devices having a high GB product, and possible external compensation (Section 3.1), extreme care in design and construction is required in order to obtain satisfactory operation; this program cannot guarantee this.

## 3.5 Exercises

**(3.1)** Write a program which enables the error in the straight line approximations of Figure 3.2 to be determined. Check that the correct error is predicted at $f = f_o$ and investigate other frequencies such as $f = 0.1, 0.5, 2$ and $10 f_o$.

**(3.2)** Extend the program of Exercise (3.1) to cover closed loop gain.

**(3.3)** Using any graph plotting facilities available on your computer, write a program which displays the Bode plots of Figures 3.2 and 3.3.

**(3.4)** Extend program P(3.2) to accept other definitions of bandwidth such as 0.1 and 1 per cent error.

**(3.5)** Modify program P(3.1) so that, for a given slew rate or full power bandwidth, the maximum possible amplitude at a specified frequency is determined.

**(3.6)** Extend program P(3.2) to permit amplifier selection on a slew rate/full power bandwidth basis.

**(3.7)** Extend program P(3.2) to detect devices of equal specification and to print their details as acceptable alternatives.

## 3.6 Reference

3.1 AROTSKY, J., TAYLOR, J. and GLASSBROOK, D. W. *Introduction to Microcomputing with the PET*. Edward Arnold 1983, pp 113–16

# Chapter 4
# Offset errors

## 4.1 Offset voltage, bias and difference currents

Operation of the basic amplifier configurations, discussed in Chapter 2, was based essentially on three assumptions:

(1) The open loop gain $A$ is infinite.
(2) The output voltage $V_o$ is zero when the net sum of the inputs is zero.
(3) The current flowing into, or out of, the amplifier inputs is zero.

For a real amplifier none of these will be precisely true. The effect of a finite value of $A$ may readily be determined and corrected (see Exercise 2.5, Section 2.7) and so will not be considered further.

Effect (2) may be allowed for by means of a hypothetical offset voltage $V_{os}$ which is normally referred to the amplifier inputs; this is shown, together with the so-called input bias currents (effect (3)) in Figure 4.1. A compensating resistor $R_c$ has also been included and will be discussed later.

Summing currents at the inverting amplifier input:

$$\frac{V_i - v_-}{R_i} + \frac{V_o - v_-}{R_f} + i_{b-} = 0$$

If $A$ is very large $v_- \simeq v_+ + v_{os}$ since it is now $(v_+ - v_- + v_{os})$ which tends to zero.

Substituting and rearranging:

$$V_o = -\left[V_i\frac{R_f}{R_i} - v_{os}\left(1 + \frac{R_f}{R_i}\right) + R_f i_{b-} - v_+\left(1 + \frac{R_f}{R_i}\right)\right] \quad (4.1)$$

The first of these terms is the required one, the second is the output error due to the offset voltage, and the third is the output error due to input bias current.

The last term is due to the presence of the compensating resistor $R_c$ and is zero if $R_c$ is zero. However, this term can be used to advantage as follows.

## Offset errors

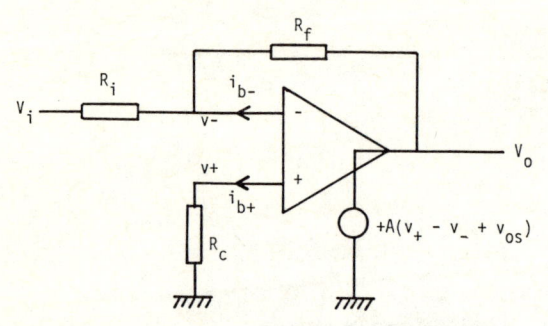

*Figure 4.1* Operational amplifier circuit showing input offset voltage and bias currents

Since $v_+ = i_{b+} R_c$, the last two terms of Equation (4.1) become

$$R_f i_b - R_c \left(1 + \frac{R_f}{R_i}\right) i_{b+}$$

If $R_f = R_c(1 + R_f/R_i)$, these two terms become

$$R_f(i_{b-} - i_{b+}) = R_f i_d$$

where $i_d$ is the (differential) input offset current. Since $i_d$ is typically around one order of magnitude smaller than $i_{b+}$ and $i_{b-}$, this reduction in offset justifies the use of an additional resistor. From the above, its value is given by

$$R_c = \frac{R_f R_i}{R_i + R_f} \tag{4.2}$$

which is the value of $R_f$ and $R_i$ in parallel. This means that the two amplifier inputs 'see' the same source resistance and only the difference between the two bias currents ($i_d$) contributes to the offset at the output.

Program P(4.1) calculates (line 130) the total error due to the offset voltage and bias current, using Equation (4.1) for the uncompensated case (that is, the term involving $v_+$ is omitted). The magnitudes of the terms are added in order to obtain a 'worst case' value; it is possible that the terms could be of opposite signs and, therefore, partially cancel. Subsequently (line 150), the required value of compensating resistor is determined using Equation (4.2) and, for a specified input offset current (line 180), the total offset is recalculated for the compensated case (line 190).

The reduction in offset due to the use of a compensating resistance is clearly shown and its usefulness may readily be assessed for a particular application (refer to the specimen runs). Notice that in the

```
10 REM PROGRAM P(4.1)
20 REM PROGRAM TO DETERMINE OFFSET ERRORS
30 REM AND TO DETERMINE COMPENSATING RESISTANCE VALUE
40 PRINT""
50 PRINT"FEEDBACK RESISTANCE VALUE (KOHMS)"
60 INPUT RF
70 PRINT"INPUT RESISTANCE VALUE (KOHMS)"
80 INPUT RI
90 PRINT"OP AMP OFFSET VOLTAGE (MILLIVOLTS)"
100 INPUT VO
110 PRINT"OP AMP BIAS CURRENT (NANOAMPS)"
120 INPUT IB
130 E=VO*(1+RF/RI)+RF*IB*1E-03
140 PRINT"TOTAL OUTPUT ERROR="E"MILLIVOLTS"
150 RC=(RF*RI)/(RF+RI)
160 PRINT "REQUIRED COMPENSATING RESISTANCE="RC"KOHMS"
170 PRINT"OP AMP INPUT OFFSET CURRENT (NANOAMPS)"
180 INPUT ID
190 EC=VO*(1+RF/RI)+RF*ID*1E-03
200 PRINT"COMPENSATED OUTPUT ERROR="EC"MILLIVOLTS"
```

*Specimen run*
```
FEEDBACK RESISTANCE VALUE (KOHMS)
? 100
INPUT RESISTANCE VALUE (KOHMS)
? 10
OP AMP OFFSET VOLTAGE (MILLIVOLTS)
? 1
OP AMP BIAS CURRENT (NANOAMPS)
? 80
TOTAL OUTPUT ERROR= 19 MILLIVOLTS
REQUIRED COMPENSATING RESISTANCE= 9.09090909 KOHMS
OP AMP INPUT OFFSET CURRENT (NANOAMPS)
? 20
COMPENSATED OUTPUT ERROR= 13 MILLIVOLTS
FEEDBACK RESISTANCE VALUE (KOHMS)
? 1000
INPUT RESISTANCE VALUE (KOHMS)
? 10
OP AMP OFFSET VOLTAGE (MILLIVOLTS)
? 1
OP AMP BIAS CURRENT (NANOAMPS)
? 80
TOTAL OUTPUT ERROR= 181 MILLIVOLTS
REQUIRED COMPENSATING RESISTANCE= 9.9009901 KOHMS
OP AMP INPUT OFFSET CURRENT (NANOAMPS)
? 20
COMPENSATED OUTPUT ERROR= 121 MILLIVOLTS
```

case of amplifiers that use field effect transistors in their input stages ('FET input types') the improvement obtained by using a compensating resistor may be negligible in view of the very low input bias and difference currents (see Exercise (4.2), Section 4.5).

Notice also that, since the current terms involve $R_f$ but not $R_i$, for a given gain a small value of $R_f$ should be chosen. This conflicts, however, with the requirements for high input resistance to avoid loading the previous stage, so a compromise must be made. The programs presented in this book should enable such a compromise to be readily achieved for a particular application.

## 4.2 Temperature and other effects

The offset voltages and bias currents, discussed in the previous section, have so far been assumed to be constant. In practice this is not the case. Both consist of a constant term, usually specified at 25°C, together with terms which specify the change with ambient temperature and supply voltage. Hence the offset voltage may be written:

$$v_{os} = V_{os} + \left(\frac{\delta v_{os}}{\delta T}\right)\Delta T + \left(\frac{\delta v_{os}}{\delta V_s}\right)\Delta V_s \tag{4.3}$$

and the bias current:

$$i_b = I_b + \left(\frac{\delta i_b}{\delta T}\right)\Delta T + \left(\frac{\delta i_b}{\delta V_s}\right)\Delta V_s \tag{4.4}$$

The manufacturers normally specify the temperature parameters in microvolts (or nanoamps) per °C and the effect of supply voltage as a rejection ratio in microvolts (of offset) per volt (of supply change). The last term in Equation (4.4) is not normally specified and can usually be neglected.

The constant terms in Equations (4.3) and (4.4) do not present a problem. In all but the least demanding of applications, provision is made for their effects to be cancelled. This normally involves the connection of an external trimming potentiometer to the amplifier in a manner specified by the device manufacturer. This adjusts the current flows in the amplifier so that zero output can be obtained for zero input. Alternatively, a small, adjustable, compensating signal may be added in using the configuration of Figure 2.4.

Having cancelled the initial offset errors, it is still necessary to ensure that the chosen configuration can meet the required offset tolerances for all possible operating conditions.

Program P(4.2) evaluates the worst case total offset, at the amplifier output. After requesting the relevant parameter values, the voltage contribution is calculated in line 200 and the current contribution in line 210. If required, these could be displayed separately; in the program as listed they are combined in line 220 and subsequently compared with the maximum permissible error (line

## Temperature and other effects 41

```
10 REM PROGRAM P(4.2)
20 REM PROGRAM TO EVALUATE WORST CASE
30 REM DRIFT PERFORMANCE
40 PRINT"FEEDBACK RESISTANCE (KOHM)"
50 INPUT RF
60 PRINT"INPUT RESISTANCE (KOHM)"
70 INPUT RI
80 PRINT"MAX OPERATING TEMP(C)"
90 INPUT TM
100 PRINT"MAX SUPPLY VARIATION(VOLTS)"
110 INPUT VM
120 PRINT"MAX PERMISSIBLE ERROR (MILLIVOLTS)"
130 INPUT EM
140 PRINT"OFFSET VOLTAGE TEMP COEFF (MICROVOLTS/C)
150 INPUT C1
160 PRINT"SUPPLY REJECTION(MICROV/V)"
170 INPUT C2
180 PRINT"BIAS CURRENT TEMP COEFF (NANOA/C)"
190 INPUT C3
200 VO=(C1*(TM-25)*1E-06)+(C2*VM*1E-06)
210 IB=C3*(TM-25)*1E-09
220 EO=(VO*(1+RF/RI)*1000)+(IB*RF*1E06)
230 PRINT"WORST CASE ERROR=";EO"MILLIVOLTS"
240 IFEO<=EM THEN 290
250 PRINT"CANNOT MEET REQUIRED SPEC. TRY AGAIN"
260 PRINT"PRESS ANY KEY TO REPEAT"
270 GET A$:IF A$="" THEN 270
280 RUN 20
290 PRINT"THEREFORE AMPLIFIER IS ADEQUATE":END
```

*Specimen run*

```
FEEDBACK RESISTANCE (KOHM)
? 1000
INPUT RESISTANCE (KOHM)
? 10
MAX OPERATING TEMP(C)
? 45
MAX SUPPLY VARIATION(VOLTS)
? .5
MAX PERMISSIBLE ERROR (MILLIVOLTS)
? 10
OFFSET VOLTAGE TEMP COEFF (MICROVOLTS/C)
? 5
SUPPLY REJECTION(MICROV/V)
? 30
BIAS CURRENT TEMP COEFF (NANOA/C)
? .5
WORST CASE ERROR= 21.615 MILLIVOLTS
CANNOT MEET REQUIRED SPEC. TRY AGAIN
PRESS ANY KEY TO REPEAT

FEEDBACK RESISTANCE (KOHM)
? 500
INPUT RESISTANCE (KOHM)
? 10
```

## Offset errors

```
MAX OPERATING TEMP(C)
? 40
MAX SUPPLY VARIATION(VOLTS)
? .1
MAX PERMISSIBLE ERROR (MILLIVOLTS)
? 10
OFFSET VOLTAGE TEMP COEFF (MICROVOLTS/C)
? 5
SUPPLY REJECTION(MICROV/V)
? 30
BIAS CURRENT TEMP COEFF (NANOA/C)
? .5
WORST CASE ERROR= 7.728 MILLIVOLTS
THEREFORE AMPLIFIER IS ADEQUATE
```

240). If the specification is not met the program may be re-run using either the parameters of a better amplifier or less demanding performance limits.

It has been assumed for simplicity that the maximum operating temperature will exceed 25°C and will define the temperature performance limit. It would be preferable to request upper and lower operating temperature limits and to use the (modulus of the) maximum deviation in lines 200 and 210. (Exercise (4.3), Section 4.5).

The specimen run shown for program P(4.2) is based on the characteristics of the standard 741 type amplifier. Initially, the demanded specification is too ambitious and cannot be met. In the re-run, the required gain is halved (by halving the feedback resistance in order to reduce bias current effects), the maximum operating temperature is reduced by 5°C and a more stable power supply is used. For the same permissible error (10 mv) the 741 is now just adequate.

### 4.3 Use of T network to reduce feedback resistance

In order to retain a high value of input resistance when operating at high gain, an impractically high value of feedback resistance can be required. Also, the bias current contribution to the total offset is directly proportional to the feedback resistance (Equation (4.1)). It is, therefore, desirable to reduce the value of feedback resistance while retaining the required gain.

This can be achieved by including an attenuator circuit, usually in the form of a T network, in the amplifier feedback path as shown in Figure 4.2. (The non-inverting input is grounded as usual of course.) It is convenient, but not essential, to make the two arms of the network ($R_t$) of equal value. Since the virtual earth point presents a very low resistance to ground, $V_o'$ may be calculated from $V_o$ by way of

Use of T network to reduce feedback resistance 43

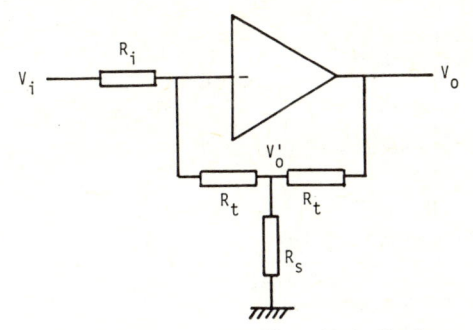

Figure 4.2 Operational amplifier with feedback attenuator

a potentiometer network consisting of $R_t$ together with $R_t$ and $R_s$ in parallel. Hence,

$$V'_o = V_o \frac{R_s R_t}{R_s R_t + R_t(R_s + R_t)} \qquad (4.5)$$

Summing currents at the amplifier input in the usual way gives:

$$\frac{V_i}{R_i} + \frac{V'_o}{R_t} = 0 \qquad (4.6)$$

Eliminating $V'_o$ from (4.5) and (4.6) gives:

$$V_o = -\frac{V_i}{R_i}\left(\frac{R_t^2 + 2R_s R_t}{R_s}\right) \qquad (4.7)$$

The term in brackets can be regarded as an effective feedback resistance; let this be $R_f(\text{eff})$. From Equation (4.7):

$$R_s = \frac{R_t^2}{R_f(\text{eff}) - 2R_t} \qquad (4.8)$$

$R_f(\text{eff})$ is the value of feedback resistance which would be required in the normal operational amplifier configuration.

Program P(4.3) evaluates Equation (4.8) for the required value of $R_f(\text{eff})$ (notice that megohms are used since large values can be expected) and an arbitrary choice of $R_t$ (lines 90 and 100). Obviously, the chosen value of $R_t$ should be significantly smaller than $R_f(\text{eff})$; if it is not less than half this value, Equation (4.8) will fail. It may be necessary to try several values of $R_t$ in order to obtain a convenient value for $R_s$.

The amplifier input bias current will now flow through the feedback network which, from Figure 4.2, can be seen to present a

## 44 Offset errors

```
10 REM PROGRAM P(4.3)
20 REM PROGRAM TO DETERMINE T-NETWORK
30 REM FEEDBACK COMPONENT VALUES
40 REM AND TO COMPARE OFFSET DUE TO BIAS CURRENT
50 REM WITH THE SINGLE FEEDBACK RESISTANCE CASE
60 PRINT""
70 PRINT"REQUIRED EFFECTIVE FEEDBACK RESISTANCE (MOHM)"
80 INPUT RE
90 PRINT"CHOOSE A VALUE FOR RT (KOHMS)"
100 INPUT RT
110 RS=((RT*1E03)↑2)/(RE*1E06-(2*RT*1E03))
120 PRINT"RS=";RS"OHMS
130 RB=(RT*1E03)+(RS*RT*1E03)/(RS+(RT*1E03))
140 PRINT"AMPLIFIER BIAS CURRENT (NANOAMPS)"
150 INPUT IB
155 PRINT "ERROR DUE TO BIAS CURRENT WITH T-NETWORK"
160 PRINT"="IB*RB*1E-06"MILLIVOLTS"
165 PRINT "ERROR DUE TO BIAS CURRENT WITH "RE"MEGOHM"
170 PRINT"FEEDBACK RESISTOR="IB*RE"MILLIVOLTS"
```

*Specimen run*

```
REQUIRED EFFECTIVE FEEDBACK RESISTANCE (MOHM)
? 100
CHOOSE A VALUE FOR RT (KOHMS)
? 47
RS= 22.1107841 OHMS
AMPLIFIER BIAS CURRENT (NANOAMPS)
? 80
ERROR DUE TO BIAS CURRENT WITH T-NETWORK
= 3.76176803 MILLIVOLTS
ERROR DUE TO BIAS CURRENT WITH  100 MEGOHM
FEEDBACK RESISTOR= 8000 MILLIVOLTS
```

source resistance of $R_t + (R_s$ and $R_t$ in parallel). The value of $R_s$ is usually small so this approximates to $R_t$. The output error due to bias current flowing in this resistance is evaluated in line 160 and compared with that due to a single feedback resistance (line 170).

The values chosen for the specimen run (using a 741) emphasize the improvement that can be obtained.

### 4.4 Blocking of D.C. offset

Operation at very low frequencies, including D.C., is an important and useful feature of operational amplifiers. However, their low cost and ease of use make them attractive for applications where very low frequency performance is not required (for example in audio amplifiers). In such cases, the effect of any offset may be eliminated by means of a capacitor, a 'blocking capacitor', which could be connected directly in series with the amplifier output. This is shown, for a two-stage amplifier, in Figure 4.3.

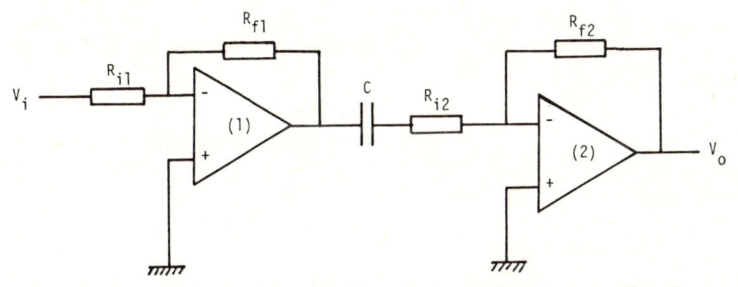

*Figure 4.3* Two-stage operational amplifier circuit with blocking capacitor $C$

For $V_i = 0$, capacitor $C$ will charge to the output offset voltage of amplifier (1) via $R_{i2}$ and the virtual earth of amplifier (2). For subsequent, non-zero, values of $V_i$ the offset voltage stored on $C$ will be effectively subtracted from the output of amplifier (1) thereby providing amplifier (2) with an input which has no D.C. component.

In multi-stage amplifiers without blocking capacitors the offset of the early stages is amplified by the gain of all subsequent stages. This can clearly lead to saturation of the later stages when high gains are required. Blocking capacitors avoid this problem by restricting the effect of offset to individual stages.

Two precautions must be observed in designing amplifiers of this kind:

(1) The offset, even within a single stage, must not be so great that saturation can occur with the peak values of the anticipated alternating signal. Programs P(4.1) and P(4.2) can be used to check for this.

(2) The blocking capacitors must have sufficiently high values that the lowest required signal frequency is not attenuated excessively. The second stage of Figure 4.3 has effectively become the high pass filter of Figure 6.3, which will be discussed later.

More sophisticated operational amplifier circuits have been developed for A.C. amplification but are beyond the scope of this book. See, for example, Reference 4.1.

## 4.5 Exercises

**(4.1)** Extend the theory of Section 4.1 and program P(4.1) to cover the determination of the required compensating resistance, and resulting offset errors, for an operational summer using several input resistances.

**(4.2)** The widely used 741 operational amplifier has an offset voltage of 1 mV, an input bias current of 80 nA and an input difference current of 20 nA. The FET input version (3140) has values of 5 mV, 10 pA and 0.5 pA respectively (notice the current units). By running program P(4.1) for some typical values of input and feedback resistance, show that a compensating resistor has negligible effect in the case of the FET input device.

**(4.3)** Modify program P(4.2) to accept upper and lower operating temperatures and to use the greater deviation from 25°C in calculating the offsets.

**(4.4)** Extend program P(4.2) using the methods of P(4.1) to allow for the improved performance which can be obtained by using a compensating resistor.

**(4.5)** Extend program P(4.2) using a 'library' of device parameters (see program P(3.2)) to select the lowest cost device which meets the required specifications from those available.

**(4.6)** Extend program P(4.3) to include voltage offset terms in order to indicate the total error.

**(4.7)** Extend program P(4.3) to check that the proposed value for $R_t$ really is significantly less than $\frac{1}{2}(R_f(\text{eff}))$.

## 4.6 Reference

4.1 GRAEME, J. G., TOBEY, G. E. and HUELSMAN, L. P. *Operational Amplifiers, Design and Applications.* McGraw-Hill 1971, pp 222–25

Chapter 5
# Waveform generation

## 5.1 Preliminary comments

Analogue integrated circuits are particularly suitable for the generation of a wide range of waveforms. Several excellent devices are available specifically for the generation of timing pulses, triangular waves, sine waves and suchlike; further details are available in the application notes of the appropriate manufacturers.

In this book, attention will be restricted to the use of operational amplifiers for the purpose of waveform generation. Before undertaking a detailed design for a specific application it would be wise to check whether a ready-made device is available at a reasonable cost.

Although a wide range of different generation techniques is available, they may be broadly classified into those based on the generation of linear ramps (which may subsequently be shaped into sinusoidal or other form by a suitable non-linear circuit) and those which generate sinusoids directly. The former are appropriate to applications where timing is the basic requirement and the latter to precision sine wave generation since harmonic distortion is difficult to remove from shaped triangular waves. These approaches will be discussed with examples in the following sections.

## 5.2 Ramp-based generators

Generators of this kind depend on the use of one or more operational amplifiers which are arranged to integrate an input voltage with respect to time. A constant input voltage, therefore, produces a linear ramp output. The required configuration is readily obtained by replacing the feedback resistor of a conventional inverting amplifier with a capacitor, as shown in Figure 5.1.

This circuit may be analysed, as previously, by summing the currents at the amplifier input and neglecting current flow into the amplifier itself, that is: $i_R + i_c = 0$. For the capacitor, $Q = CV$, where $Q$ is the charge stored and $V$ the voltage across the capacitor

47

## 48 Waveform generation

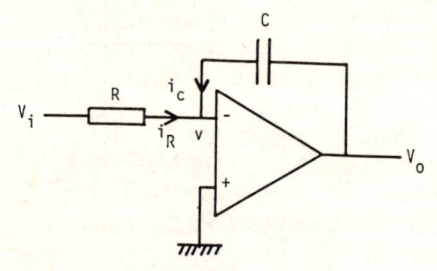

*Figure 5.1* Operational integrator configuration

$(= V_o - v)$. Therefore

$$i_c = \frac{dQ}{dt} = C\frac{d(V_o - v)}{dt}$$

and

$$i_R = \frac{V_i - v}{R}$$

Therefore

$$\frac{V_i - v}{R} + C\frac{d(V_o - v)}{dt} = 0$$

If the open loop gain $A$ of the amplifier is large, $v$ will tend to zero, hence:

$$\frac{V_i}{R} = -C\frac{dV_o}{dt}$$

Integrating both sides gives the output voltage at time $T$:

$$V_o = -\frac{1}{RC}\int_0^T V_i \, dt + (V_o)_o \qquad (5.1)$$

$(V_o)_o$ is the arbitrary constant of integration; in practice it represents the initial charge on the capacitor (i.e. for $t = 0$). It can be made zero, if required, by temporarily switching a low value resistance in parallel with the capacitor before integration starts.

The term $(1/RC)$ is the reciprocal of the resistor capacitor time constant and is often referred to as the 'gain' of the integrator. However, unlike conventional amplifier gain which is dimensionless, it has the dimensions of inverse time and is often quoted in 'volts per second per volt' which emphasizes the fact that a constant input produces a constant rate of change of voltage at the output. It is also important to note that, if the amplifier has *any* uncompensated offset error, there will be a rate of change at the output even for zero input voltage (see Exercise 5.1). This means that the amplifier will eventually saturate and so cannot be used in isolation without some arrangement to ensure that it remains within its proper operating region.

In order to avoid this difficulty and generate a repetitive waveform, one or more comparators (see Section 2.1) are required in order to detect when the integrator output has reached the required upper and lower levels. When these critical levels have been reached, the sign of the integrator input voltage must be reversed, thereby reversing the rate of change of its output. Unfortunately this simple arrangement is not satisfactory since, as soon as the integrator output has reversed, the comparator will revert to its previous state and change the direction of integration yet again. The resulting triangular wave is of negligibly small amplitude!

The solution is a comparator with relatively large hysteresis arranged so that this defines the amplitude of the generated waveform. Fortunately, a comparator with hysteresis may be simply configured as shown in Figure 5.2. This looks very much like a conventional operational amplifier configuration, but notice that the feedback is applied to the *non-inverting* input. In view of the resulting positive feedback, $V_o$ will always be at either the upper or lower saturation voltage (Figure 2.2). Assume initially that $V_i$ and $V_o$ are both positive; clearly, the non-inverting input of the amplifier will be positive and $V_o$ is held at its positive saturation value. Only when $V_o$ is taken sufficiently negative in order to absorb all the current flowing through $R_f$ will the non-inverting input be taken negative thereby causing the amplifier to 'flip over' into its negative saturation state. A subsequent similar reversal will take place when $V_i$ is taken sufficiently positive. This change of state will clearly occur when

$$\frac{V_i}{R_i} = -\frac{(V_o)_{sat}}{R_f} \qquad (5.2)$$

where sat denotes the saturation value.

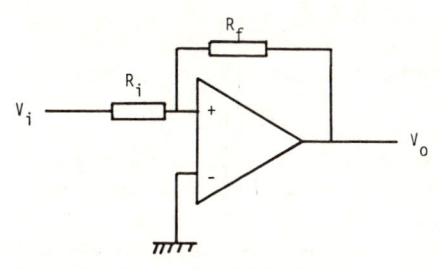

*Figure 5.2* Comparator with hysteresis

Normally, $R_f$ will be made greater than $R_i$ so that $V_i$ is less than $(V_o)_{sat}$ and the comparator can be switched by a preceding amplifier with similar output saturation voltage. If $R_f = nR_i$, the switching levels become $\pm(V_o)_{sat}/n$ as shown in Figure 5.3.

The comparator with hysteresis can readily be combined with an operational integrator in order to provide a triangular and square wave generator as shown in Figure 5.4. $V_T$ will be a triangular wave with nearly equal positive and negative slopes and $V_s$ will be a square wave which is positive during the falling ramp of $V_T$, and vice versa. Matching of the positive and negative saturation voltages, and hence positive and negative slopes of $V_T$, can be improved by using Zener diodes and a current limiting resistor at the output of the comparator (Ref. 5.1). The limiting voltages become the operating voltages of the Zener diodes. These are stable to within a few per cent whereas amplifier saturation voltages can vary markedly. The overall D.C. level of the waveform may be adjusted by returning the inverting

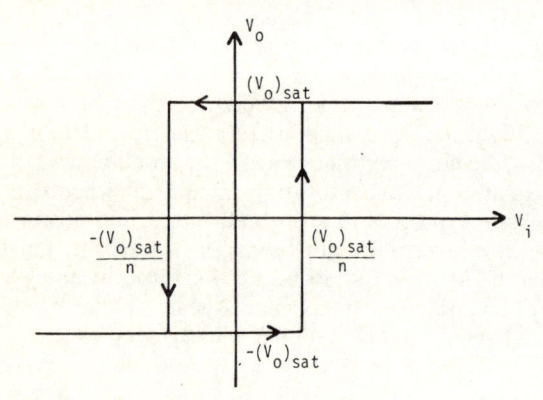

*Figure 5.3* Characteristic of comparator with hysteresis

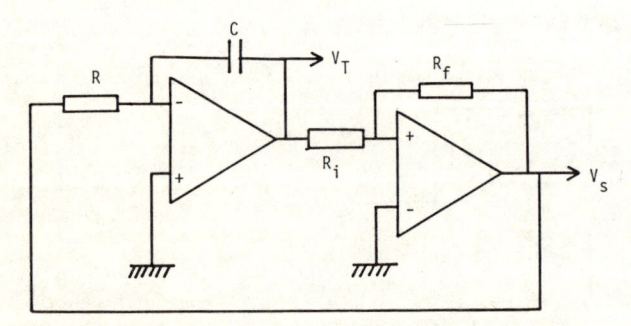

*Figure 5.4* Basic triangular waveform generator

Sine wave oscillators 51

input of the comparator amplifier to an appropriate variable voltage. Different gradients for the rising and falling ramps may be obtained by means of diodes which select different integrator input resistances in the two cases (Ref. 5.2).

In particular, the falling ramp may be made very fast, giving an approximation to a sawtooth waveform, by using a high-speed switching device in order to discharge the integrating capacitor rapidly. The circuit described in Reference 5.3 uses a programmable unijunction transistor for this purpose.

Program P(5.1) accepts the ratio of $R_f$ to $R_i$ for the comparator (line 50) and checks that this is greater than 1 (line 60). After accepting the required frequency (lines 90 and 100) the period and required time constant are calculated (lines 120 and 130). The operator suggests a value for the integrating capacitor (line 160) and the required integrator input resistance is printed (line 180). If this is not a convenient value (lines 190 and 220) a new value of integrating capacitor can be tried. The various possibilities are shown in the specimen runs.

## 5.3 Sine wave oscillators

Circuits in this category are characterized by positive feedback giving a closed loop gain very close to unity and a frequency sensitive network which ensures that the required conditions for oscillation are

```
10 REM PROGRAM P(5.1)
20 REM PROGRAM TO DETERMINE COMPONENT VALUES
30 REM FOR TRIANGULAR WAVE OSCILLATOR
40 PRINT"RATIO OF RF TO RI"
50 INPUT N
60 IF N > 1 THEN 90
70 PRINT"N MUST BE GREATER THAN 1"
80 GOTO 40
90 PRINT"REQUIRED FREQUENCY (HZ)"
100 INPUT F
110 REM CALCULATE PERIOD (T)
115 REM AND REQUIRED TIME CONSTANT (TC)
120 T=1/F
130 TC = N*T/4
140 PRINT "REQUIRED TIME CONSTANT = "TC"SECONDS"
150 PRINT"INTEGRATING CAPACITOR VALUE (MICROFARADS)"
160 INPUT C
170 R = (TC/C)*1000
180 PRINT"INTEGRATOR INPUT RESISTANCE = "R"KILOHMS"
190 PRINT"IF THIS IS NOT CONVENIENT TYPE N"
200 PRINT"PRESS ANY OTHER KEY TO FINISH"
210 GET A$:IF A$ = "" THEN 210
220 IF A$ ="N"THEN 150
230 END
```

## 52 Waveform generation

### Specimen run

```
RATIO OF RF TO RI
? .2
N MUST BE GREATER THAN 1
RATIO OF RF TO RI
? 2
REQUIRED FREQUENCY (HZ)
? 50
REQUIRED TIME CONSTANT =  .01 SECONDS
INTEGRATING CAPACITOR VALUE (MICROFARADS)
? .1
INTEGRATOR INPUT RESISTANCE =  100 KILOHMS
IF THIS IS NOT CONVENIENT TYPE N
PRESS ANY OTHER KEY TO FINISH

RATIO OF RF TO RI
? 3
REQUIRED FREQUENCY (HZ)
? 1000
REQUIRED TIME CONSTANT =  7.5E-04 SECONDS
INTEGRATING CAPACITOR VALUE (MICROFARADS)
? .1
INTEGRATOR INPUT RESISTANCE =  7.5 KILOHMS
IF THIS IS NOT CONVENIENT TYPE N
PRESS ANY OTHER KEY TO FINISH
N
INTEGRATING CAPACITOR VALUE (MICROFARADS)
? .015
INTEGRATOR INPUT RESISTANCE =  50 KILOHMS
IF THIS IS NOT CONVENIENT TYPE N
PRESS ANY OTHER KEY TO FINISH
```

satisfied only at the required frequency. Many configurations have been used; only the Wien bridge based oscillator, which is widely used, will be discussed.

The basic circuit is shown in Figure 5.5 where $R_1 C_1 R_2 C_2$ constitute the Wien bridge and the amplifier provides an adjustable positive gain. The output from the filter $V_f$ will be related to the output of the amplifier $V_o$ by

$$V_f = V_o \frac{Z_2}{Z_1 + Z_2}$$

where $Z_1$ is the impedance of $R_1$ and $C_1$ in series and $Z_2$ is the impedance of $R_2$ and $C_2$ in parallel; that is:

$$Z_1 = R_1 + 1/j\omega C_1$$

$$Z_2 = \frac{R_2(1/j\omega C_2)}{R_2 + 1/j\omega C_2} = \frac{R_2}{1 + j\omega C_2 R_2}$$

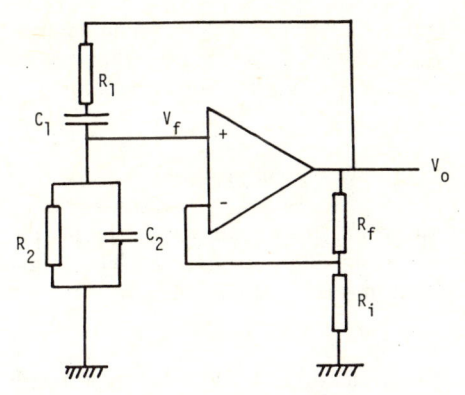

Figure 5.5 Basic Wien bridge oscillator

Substituting and simplifying gives

$$V_f = V_o \frac{R_2}{(R_1 + R_2 + R_2C_2/C_1) + j(\omega R_1 R_2 C_2 - 1/\omega C_1)}$$

If $R_1 = R_2 = R$ and $C_1 = C_2 = C$, which is convenient but not essential:

$$V_f = V_o \frac{R}{3R + j(\omega R^2 C - 1/\omega C)} \tag{5.3}$$

For oscillations to be maintained there must be no net phase shift so the imaginary part of Equation (5.3) will be zero at the frequency of oscillation; hence

$$\omega = 1/RC \quad \text{or} \quad f = 1/2\pi RC \tag{5.4}$$

At this frequency the gain of the network will be 1/3 so the amplifier must provide a gain of 3 for oscillations to be maintained. From Equation (2.6), $(R_f + R_i)/R_i = 3$ and hence $R_f = 2R_i$.

Unfortunately, this basic arrangement is not satisfactory since the required loop gain of *precisely* 1 cannot be maintained. Even a small decrease means that the oscillations die away, and any increase causes the oscillations to build up until saturation of the amplifier causes severe distortion. Many solutions to this problem have been suggested; all involve a non-linear element of some kind, which ensures a gain slightly in excess of unity at low levels of oscillation and which falls below this value as the required amplitude of oscillation is exceeded. This ensures that oscillations build up when the circuit is first switched on and stabilize at the required amplitude.

One possibility (Ref. 5.4) is the use of back-to-back Zener diodies

## 54 Waveform generation

connected across $R_f$ of Figure 5.5 in conjunction with an adjustable value of $R_i$. The latter is set for a loop gain slightly greater than unity. When the amplitude of oscillation exceeds the Zener voltage the diodes begin to conduct and hence reduce the effective value of $R_f$ and consequently the loop gain.

Unfortunately, any non-linear network will cause some distortion of the generated sine wave but, with careful design, this can be made small. Reference 5.4 also describes an improved amplitude stabilizing circuit which incorporates a field effect transistor. The high impedance input of this device minimizes the clipping effect on the generated waveform.

Program P(5.2) enables the required component values for a specified frequency of oscillation (lines 40 and 50) to be determined using Equation (5.3). Either R or C must be chosen by the operator; a choice is provided in lines 60 to 90. The appropriate routine is then entered and the required component value printed out (line 140 or 230). In both cases the calculation can be repeated using a more appropriate initial value. The specimen runs show some typical cases.

```
10 REM PROGRAM P(5.2)
20 REM PROGRAM TO DETERMINE COMPONENT VALUES
30 REM FOR WIEN BRIDGE OSCILLATOR
40 PRINT"ENTER REQUIRED FREQUENCY (HZ)"
50 INPUT F
60 PRINT"CHOOSE RESISTANCE (R) OR CAPACITANCE (C)?"
70 GET A$:IF A$="" THEN 70
80 IF A$="R" THEN 110
90 IF A$="C" THEN 200
100 GOTO 70
110 PRINT "ENTER RESISTANCE VALUE (KILOHMS)"
120 INPUT R
130 C = 1000/(2*π*F*R)
140 PRINT "REQUIRED CAPACITANCE = "C"MICROFARADS"
150 PRINT"TRY A DIFFERENT RESISTANCE (Y/N)?"
160 GET A$:IF A$ = "" THEN 160
170 IF A$ = "Y" THEN 110
180 IF A$ = "N" THEN 290
190 GOTO 160
200 PRINT"ENTER CAPACITANCE VALUE (MICROFARADS)"
210 INPUT C
220 R = 1000/(2*π*F*C)
230 PRINT"REQUIRED RESISTANCE = "R"KILOHMS"
240 PRINT "TRY A DIFFERENT CAPACITANCE (Y/N)?"
250 GET A$:IF A$ = ""THEN 250
260 IF A$ = "Y" THEN 200
270 IF A$ = "N" THEN 290
280 GOTO 250
290 END
```

## Specimen runs

1. ```
   ENTER REQUIRED FREQUENCY (HZ)
   ? 1000
   CHOOSE RESISTANCE (R) OR CAPACITANCE (C)?
   R
   ENTER RESISTANCE VALUE (KILOHMS)
   ? 22
   REQUIRED CAPACITANCE =  7.2343156E-03 MICROFARADS
   TRY A DIFFERENT RESISTANCE (Y/N)?
   N
   ```

2. ```
   ENTER REQUIRED FREQUENCY (HZ)
   ? 1000
   CHOOSE RESISTANCE (R) OR CAPACITANCE (C)?
   C
   ENTER CAPACITANCE VALUE (MICROFARADS)
   ? 4.7E-03
   REQUIRED RESISTANCE =   33.8627539 KILOHMS
   TRY A DIFFERENT CAPACITANCE (Y/N)?
   N
   ```

3. ```
   ENTER REQUIRED FREQUENCY (HZ)
   ? 50
   CHOOSE RESISTANCE (R) OR CAPACITANCE (C)?
   R
   ENTER RESISTANCE VALUE (KILOHMS)
   ? 27
   REQUIRED CAPACITANCE =  .11789255 MICROFARADS
   TRY A DIFFERENT RESISTANCE (Y/N)?
   Y
   ENTER RESISTANCE VALUE (KILOHMS)
   ? 330
   REQUIRED CAPACITANCE =  9.64575413E-03 MICROFARADS
   TRY A DIFFERENT RESISTANCE (Y/N)?
   N
   ```

## 5.4 Exercises

**(5.1)** Modify program P(4.1) so that it computes the rate output for an operational integrator with an offset voltage and input bias current.

**(5.2)** Extend program P(5.1) so that either a required frequency or period can be accepted.

**(5.3)** Extend program P(5.2) to cover the case where $R_1 \neq R_2$ and $C_1 \neq C_2$.

## 5.5 References

5.1 STOUT, D. F. and KAUFMAN, M. (Eds) *Handbook of Operational Amplifier Circuit Design*. McGraw-Hill 1976, page 5-13
   ```

5.2 As 5.1, page 27-4
5.3 COUGHLIN, R. F. and DRISCOLL, F. F. *Operational Amplifiers and Linear Integrated Circuits.* Prentice Hall 1982, pp 124–5
5.4 JACOB, J. M. *Applications and Design with Analog Integrated Circuits.* Reston 1982, pp 287–9

Chapter 6
# Introduction to active filters

## 6.1 Preliminary comments

Filters are networks whose gain (or attenuation) and associated phase shift vary with frequency. This variation can be exploited in order to separate wanted and unwanted components of a signal on the basis of frequency. In general, filters consist of linear elements such as resistors, capacitors and inductors although useful filters can be constructed using any two of these. Inductors are widely used at radio frequencies where the required values are a convenient size. At lower frequencies resistance-capacitance filters predominate in view of the bulk, cost and non-linearity of iron-cored inductors which would be required to produce sufficiently high values of inductance.

Filters may be broadly classified as passive or active. The former use resistive and reactive components alone and cannot, therefore, produce power gain, but active filters use these components in conjunction with some form of amplifying circuit and can, therefore, produce power gain. Operational amplifiers are particularly suitable for this application.

Passive filters have the advantages of physical simplicity, operation at almost any required frequency and lack of power supply requirements. A disadvantage is that source and load impedances must be taken into account when calculating the characteristics of the filter since these effectively form part of the network. If they change, the filter characteristics may change markedly. Also, component interactions mean that the filter must be analysed as a whole, which often leads to involved mathematics. Active filters, on the other hand, can use buffer amplifiers in order to isolate each section of the filter from source and load impedance effects. Buffers between stages of filters mean that transfer functions can be cascaded by simple multiplication; this simplification is clearly shown in Figure 6.1.

Provided the voltage followers are ideal,

$$v_o = v_s \frac{1}{(1 + jC_1 R_1)(1 + jC_2 R_2)} \tag{6.1}$$

58   Introduction to active filters

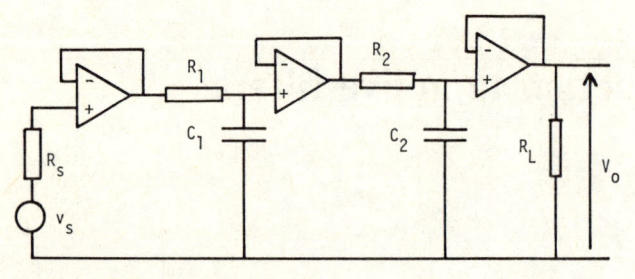

*Figure 6.1* Cascaded filter networks

where the terms in brackets are the transfer functions of the individual filter networks (compare Equation (3.1)). Notice that $R_s$ and $R_L$ do not appear in this expression.

A further advantage of active filters is that the high input impedance of amplifiers means that long time constants can be obtained by using high resistance values without the need for very large capacitance values. Of course, active filters cannot operate at frequencies where the constituent amplifiers no longer conform sufficiently accurately to the assumptions of ideal operation. For these reasons, active filters tend to be used at relatively low frequencies with passive circuits being used at higher frequencies.

Filters may also be classified by their 'order'. This is determined essentially by the number of reactive components in the filter, which in turn determines the order of the equation that relates the output of the filter to its input (the 'transfer function'). The example of Figure 6.1 is, therefore, a second order filter realized, in this case, by cascading two first order filters.

It has already been shown (Section 3.1) that the gain of a first order filter falls off at 20 dB per decade regardless of the detailed configuration. In general, the gain of an $n$th order filter will fall off at $20n$ dB per decade. (The fall off will be at low frequencies for high pass filters and vice versa.)

The minimum order of filter required to separate two components of different frequency by a specified amount can, therefore, readily be determined. This is achieved by program P(6.1).

The two frequencies, together with the maximum acceptable attenuation of the wanted component and the minimum acceptable rejection of the unwanted component are obtained in lines 50 to 120. The frequency ratio is calculated and made positive if necessary (lines 130 and 140). The minimum required gain difference is calculated in line 150 and the attenuation of a *first order* filter for the particular frequency ratio is determined (line 160). The order of the filter is progressively increased by the FOR loop (lines 170 to 190) until the

```
10 REM PROGRAM P(6.1)
20 REM PROGRAM TO DETERMINE ORDER OF FILTER REQIRED
30 REM FOR A SPECIFIED REJECTION
40 PRINT ""
50 PRINT "REQUIRED FREQUENCY (HZ)"
60 INPUT FW
70 PRINT "MAX. ATTENUATION AT THIS FREQUENCY (DB)"
80 INPUT AW
90 PRINT "FREQUENCY TO BE REJECTED (HZ)"
100 INPUT FU
110 PRINT "MIN. ATTENUATION AT THIS FREQUENCY (DB)"
120 INPUT AU
130 FR = FW/FU
140 IF FR < 1 THEN FR = 1/FR
150 A = AU - AW
160 DB = 20*(LOG(FR)/LOG(10))
170 FOR I = 1 TO 10
180 IF DB*I > A THEN 230
190 NEXT I
200 PRINT "A TENTH ORDER FILTER IS INSUFFICIENT"
210 PRINT "TO REALISE THE REQUIRED SPECIFICATION"
220 END
230 PRINT "A FILTER OF ORDER "I" IS REQUIRED"
240 END
```

## Specimen runs

```
REQUIRED FREQUENCY (HZ)
? 1000
MAX. ATTENUATION AT THIS FREQUENCY (DB)
? 3
FREQUENCY TO BE REJECTED (HZ)
? 50
MIN. ATTENUATION AT THIS FREQUENCY (DB)
? 55
A FILTER OF ORDER  2  IS REQUIRED

REQUIRED FREQUENCY (HZ)
? 50
MAX. ATTENUATION AT THIS FREQUENCY (DB)
? 3
FREQUENCY TO BE REJECTED (HZ)
? 1000
MIN. ATTENUATION AT THIS FREQUENCY (DB)
? 55
A FILTER OF ORDER  2  IS REQUIRED

REQUIRED FREQUENCY (HZ)
? 1000
MAX. ATTENUATION AT THIS FREQUENCY (DB)
? 0
FREQUENCY TO BE REJECTED (HZ)
? 50
```

```
MIN. ATTENUATION AT THIS FREQUENCY (DB)
? 55
A FILTER OF ORDER  3  IS REQUIRED


REQUIRED FREQUENCY (HZ)
? 100
MAX. ATTENUATION AT THIS FREQUENCY (DB)
? 3
FREQUENCY TO BE REJECTED (HZ)
? 50
MIN. ATTENUATION AT THIS FREQUENCY (DB)
? 45
A FILTER OF ORDER  7  IS REQUIRED


REQUIRED FREQUENCY (HZ)
? 60
MAX. ATTENUATION AT THIS FREQUENCY (DB)
? 1
FREQUENCY TO BE REJECTED (HZ)
? 50
MIN. ATTENUATION AT THIS FREQUENCY (DB)
? 40
A TENTH ORDER FILTER IS INSUFFICIENT
TO REALISE THE REQUIRED SPECIFICATION
```

required attenuation is achieved and the required order is printed (line 230).

Since very high order filters are not easy to realize, the maximum order is restricted to 10 by the loop. This could, of course, be changed if required.

Several specimen runs are presented which should be self-explanatory. The final demanding specification proves to be beyond the capability of a tenth-order filter and an appropriate message is printed.

High order filters can be realized in various ways (see Reference 6.1, for example). However, in many cases, it is convenient to use cascaded second order filters (as discussed in Section 6.3) with a single first order filter (Section 6.2) if the overall required order is odd. It is important to notice that cascaded first order filters (Figure 6.1) do not provide the flexibility of specification of a true second order arrangement and therefore cannot be used as the basic 'building blocks'.

**6.2 First order active filters**

First order filters contain only one reactive component. A simple active low pass filter may be realized using one of the *RC* sections

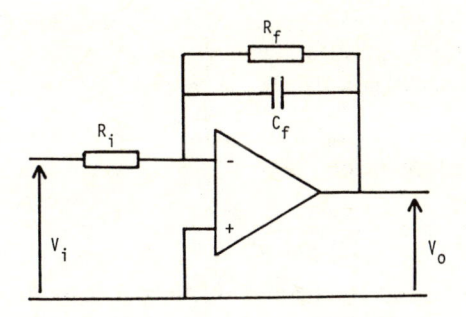

Figure 6.2 Active first order low pass filter

shown in Figure 6.1 together with input and output buffer amplifiers. Its gain/frequency response will be that of the first order model discussed in Chapter 3 and shown in Figure 3.2. The break frequency is $\omega_o = 1/RC$ and the low frequency gain tends to unity.

A more convenient arrangement is shown in Figure 6.2; this allows low frequency gains other than unity to be obtained. The analysis which led to Equation (2.3) may easily be generalized to cover impedances by writing $Z$ in place of $R$. In this case $Z_f$ becomes $R_f$ in parallel with $C_f$, and $R_i$ is unchanged. Hence:

$$Z_f = \frac{R_f 1/j\omega C_f}{R_f + 1/j\omega C_f} = \frac{R_f}{1 + j\omega C_f R_f} \tag{6.2}$$

Therefore

$$\frac{V_o}{V_i} = -\frac{Z_f}{Z_i} = -\frac{R_f}{R_i}\left(\frac{1}{1 + j\omega C_f R_f}\right) \tag{6.3}$$

This shows that the amplifier has a low frequency gain ($\omega \to 0$) defined, as usual, by $-R_f/R_i$ in conjunction with the response of a low pass filter of time constant $C_f R_f$ which does not depend on $R_i$. This implies that several inputs could be summed, with different weights if required, and the break frequency would be the same for each input.

Rationalizing Equation (6.3) and evaluating the magnitude and phase angle gives

$$\left|\frac{V_o}{V_i}\right| = \frac{R_f}{R_i}\left[\frac{1}{\sqrt{(1 + \omega^2 C_f^2 R_f^2)}}\right] \tag{6.4}$$

and

$$\angle \frac{V_o}{V_i} = -\arctan \omega R_f C_f \tag{6.5}$$

Equation (6.5) shows that the phase angle is zero for very low frequencies, tends to $-90°$ (i.e. lagging) for very high frequencies and is $-45°$ at the break frequency given by $\omega = 1/R_f C_f$ so that $\omega R_f C_f = 1$.

The configuration shown in Figure 6.3(a) enables a high pass characteristic to be obtained. As before:

$$\frac{V_o}{V_i} = -\frac{Z_f}{Z_i} = -\frac{R_f}{R_i + 1/j\omega C_i} = -\frac{j\omega C_i R_f}{1 + j\omega C_i R_i} \qquad (6.6)$$

From this it can be seen that, at very low frequencies, the gain tends to zero and, at very high frequencies, when the 1 in the denominator becomes insignificant, to $-(R_f/R_i)$.

(a)

(b)

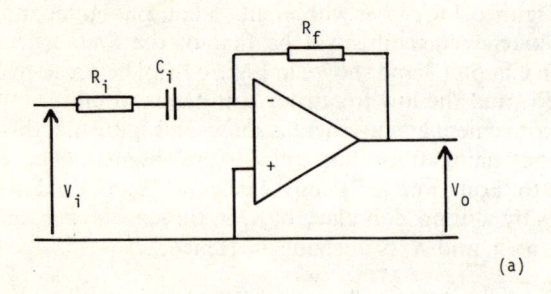

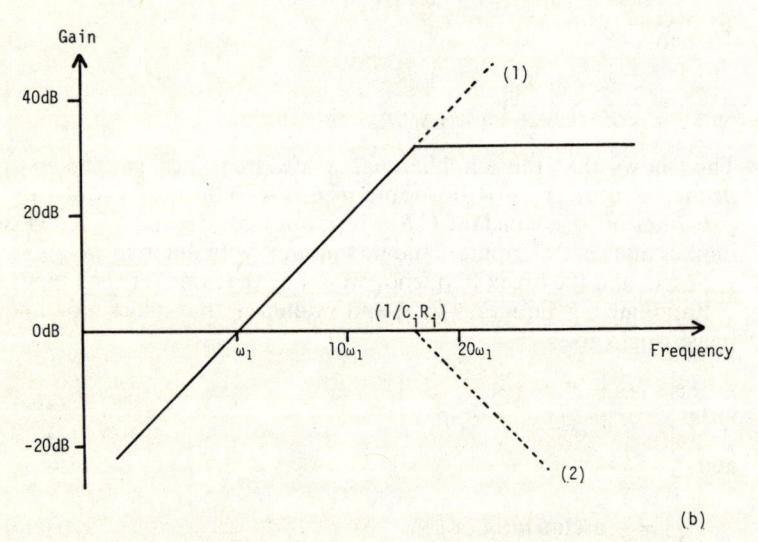

*Figure 6.3* Active first order high pass filter (a) Circuit diagram (b) Bode plot

The interpretation of Equation (6.6) is simplified by the fact that Bode plots (Section 3.1), being logarithmic, may be added when transfer functions are cascaded. The numerator represents a gain frequency characteristic with a constant rise of 20 dB per decade; it will cross the 0 dB axis when $\omega C_i R_f = 1$ (at $\omega_1$, say, where $\omega_1 = 1/C_i R_f$) as shown by curve (1) in Figure 6.3(b). The denominator represents a low pass function with break frequency given by $\omega = 1/C_i R_i$ as shown by curve (2). These two curves may be added to give the solid characteristic, shown in the figure, which is clearly a high pass one. Notice that the high frequency gain depends on the ratio of $R_f$ and $R_i$ and can be equal to, greater than, or less than unity. The phase characteristics may also be added; the numerator will contribute 90° lead at all frequencies. When combined with the low pass response, the overall effect is 90° lead at low frequencies, 45° lead at the break frequency ($\omega = 1/C_i R_i$) and zero phase shift at high frequencies.

The input arrangement of Figure 6.3(a) may be combined with the feedback arrangement of Figure 6.2 to produce a band pass filter which will have a gain that approaches $-(R_f/R_i)$ over the required range of frequencies. The gain will fall off at 20 dB per decade at both low and high frequencies. Notice that, for narrow bandwidths, if the two break frequencies are sufficiently close to each other, the gain of $(-R_f/R_i)$ will not be reached.

## 6.3 Second order active filters

As explained in Section 6.1, second order active filters, based on a single operational amplifier are useful both in their own right and as building blocks for higher order filters. Although many configurations are possible, that shown in general form in Figure 6.4 is very widely used. The components which are selected for each $Z$ determine the form of the filter response (low, high or band pass) as will be shown later. In most cases (the band pass filter is the only exception) each consists of a single resistor or capacitor as appropriate. The resistances $R_f$ and $R_i$ determine the maximum available gain and may be omitted if unity is acceptable. (In this case the amplifier output would be connected to the inverting input either directly or via a resistance equal to that 'seen' by the non-inverting input in order to reduce bias current effects (Section 4.1).)

Analysis of this circuit using standard techniques (Ref. 6.2) yields:

$$\frac{V_o}{V_i} = \frac{A_o Z_3 Z_4}{Z_1 Z_2 + Z_2 Z_3 + Z_3 Z_4 + Z_1 Z_3 + Z_1 Z_4 (1 - A_o)} \qquad (6.7)$$

where $A_o$ is the follower gain ($= 1 + R_f/R_i$).

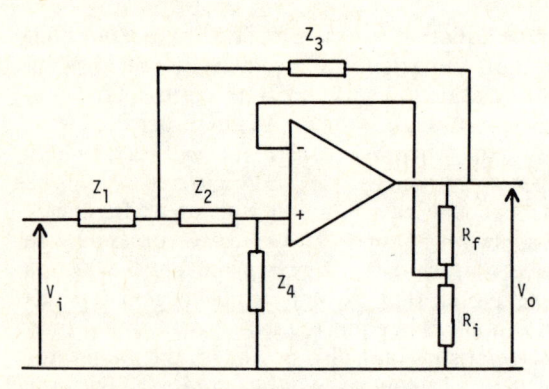

*Figure 6.4* General second order active filter

### 6.3.1 Low pass filter

To produce a low pass filter, $Z_1$ and $Z_2$ are made resistive ($R_1$ and $R_2$ respectively) and $Z_3$ and $Z_4$ are made capacitive ($C_1$ and $C_2$ respectively). Equation (6.7) therefore becomes

$$\frac{V_o}{V_i}(s) = \frac{A_o(1/sC_1)(1/sC_2)}{R_1R_2 + R_2(1/sC_1) + (1/sC_1)(1/sC_2) + R_1/(sC_1) + R_1(1-A_o)/(sC_2)}$$

where $s$ is the Laplace operator (see, for example, Reference 6.3) which may, for the purposes of this book, be regarded as equivalent to $j\omega$.

Rearranging the terms of the above equation:

$$\frac{V_o}{V_i}(s) = \frac{A_o/(R_1R_2C_1C_2)}{s^2 + s[1/(R_1C_1) + 1/(R_2C_1) + (1-A_o)/(R_2C_2)] + 1/(R_1R_2C_1C_2)}$$

(6.8)

which is of the general form

$$\frac{V_o}{V_i}(s) = A_o \frac{\omega_o^2}{s^2 + 2\zeta\omega_o s + \omega_o^2}$$

(6.9)

where $\omega_o$ is called the natural frequency and $\zeta$ is the damping factor. In this case:

$$\omega_o = 1/\sqrt{(R_1C_1R_2C_2)}$$

(6.10)

and

$$\zeta = \left(\frac{1}{R_1C_1} + \frac{1}{R_2C_1} + \frac{1-A_o}{R_2C_2}\right)\frac{\sqrt{(R_1C_1R_2C_2)}}{2}$$

$$= \frac{R_2C_2 + R_1C_2 + (1-A_o)R_1C_1}{2\sqrt{(R_1C_1R_2C_2)}} \tag{6.11}$$

This equation has been studied very widely in connection with linear dynamic systems (particularly in relation to feedback control; see, for example, Reference 6.4).

In filter theory $2\zeta$ is often written as $\alpha$ (or some other symbol) and referred to as a damping coefficient. For all values of $\alpha$, the second order low pass filter has the following characteristics.

(a) low frequency gain of $A_o$,
(b) high frequency gain which falls off at 40 dB per decade (twice the rate of a first order filter),
(c) zero phase shift at very low frequencies,
(d) 180° (lagging) phase shift at high frequencies, and
(e) 90° (lagging) phase shift at the natural frequency ($\omega_o$).

The only, but important, effect of $\alpha$ is on the nature of the gain and phase characteristics in the vicinity of $\omega_o$. For small values of $\alpha$ the response relative to very low frequencies rises before starting to fall, as shown in Figure 6.5, and the change of phase is more abrupt. The peak of the amplitude response occurs at a frequency which is somewhat less than $\omega_o$. The frequency of the peak tends towards $\omega_o$

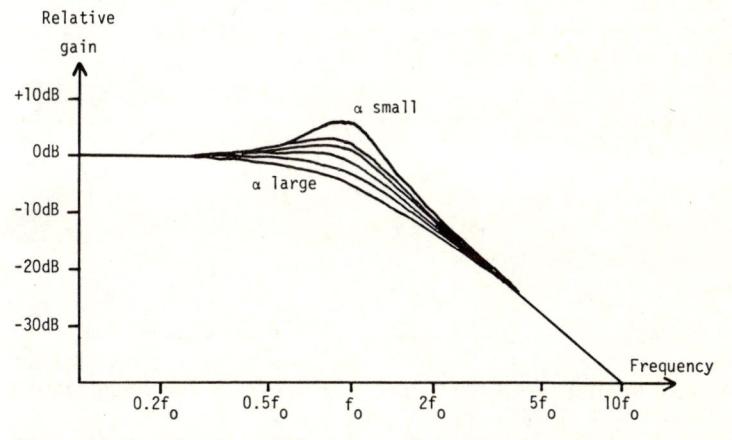

Figure 6.5 Second order gain/frequency relationships for a range of damping coefficients $\alpha$

for very small values of $\alpha$. Those values of $\alpha$ that have particular properties have been specifically named:

(1) *Bessel filter*. In some applications it is desirable that all frequency components of the signal should be delayed by an equal amount. It can be shown (Ref. 6.5) that this occurs when the variation of phase with frequency is linear. The second order Bessel filter ($\alpha = 1.73$) closely meets this requirement. However, its amplitude response starts to fall off relatively early in the pass band and its transient response is heavily damped.

(2) *Butterworth filter*. This gives the flattest amplitude response before roll-off starts to occur and is widely used for this reason. The transient response, however, can show a slight overshoot. For a second order filter the required value of $\alpha$ is $\sqrt{2}$. Since $\alpha = 2\zeta$, this corresponds to $\zeta = 1/\sqrt{2}$ which is often regarded as providing the 'best' response in linear second order control systems.

(3) *Chebyshev filters*. These have relatively low values of $\alpha$ and are characterized by a peak in the gain/frequency response followed by a rapid *initial* fall off (the final value will be 40 dB/decade as usual). The rate of initial fall off increases as the height of the peak increases; values above 3 dB are not normally used.

The required values of $\alpha$, for a second order low pass filter, are summarized in Table 6.1.

**TABLE 6.1. Damping coefficients for second order filters**

| Type of filter | Height of peak relative to D.C. (dB) | $\alpha$ |
|---|---|---|
| Bessel | — | 1.732 |
| Butterworth | — | 1.414 |
| Chebyshev | 0.1 | 1.303 |
|  | 0.25 | 1.236 |
|  | 0.5 | 1.158 |
|  | 1 | 1.046 |
|  | 2 | 0.8862 |
|  | 3 | 0.7666 |

A filter of this type will be fully specified if the natural frequency and damping factor are known (the latter defines the type of filter). For a Chebyshev type the required frequency at the peak of the response $\omega_p$ may be specified; this is *not* $\omega_o$ which may be calculated as follows (see, for example, Reference 6.6):

$$\omega_p = \omega_o \sqrt{(1 - 2\zeta^2)} = \omega_o \sqrt{(1 - \alpha^2/2)}$$

Second order active filters 67

Therefore

$$\omega_o = \omega_p \sqrt{[2/(2 - \alpha^2)]} \qquad (6.12)$$

Equation (6.8) shows that, in order to realize these two parameters, no less than six component values must be determined (since $A_o$ is determined by $R_f$ and $R_i$). Following the pioneering work of Sallen and Key (Ref. 6.7), two approaches have been developed which greatly reduce the required design effort.

*(1) Unity gain filters*
In this case $A_o$ is made equal to 1 (the amplifier operates as a simple voltage follower) and the two remaining resistors are made equal to each other ($R_1 = R_2 = R$).
From Equations (6.10) and (6.11) and since $\alpha = 2\zeta$

$$\omega_o = 1/[R\sqrt{(C_1 C_2)}] \qquad (6.13)$$

$$\alpha = 2\sqrt{(C_2/C_1)} \qquad (6.14)$$

Program P(6.2) makes use of these equations in order to determine suitable component values. The program starts by asking which type

```
10 REM PROGRAM P(6.2)
20 REM SALLEN KEY UNITY GAIN FILTER DESIGN
30 PRINT ""
40 PRINT "WHICH TYPE OF FILTER IS REQUIRED?"
50 PRINT "BESSEL(BE),BUTTERWORTH(BU) OR CHEBYSHEV(CH)"
60 INPUT F$
70 IF F$ = "BE" THEN A = 1.732:GOTO 290
80 IF F$ = "BU" THEN A = 1.414:GOTO 290
90 IF F$ = "CH" THEN 110
100 GOTO 40
110 PRINT "WHAT IS THE REQUIRED PEAK GAIN?"
120 PRINT "0.1,0.25,0.5,1,2 OR 3 DB?"
130 INPUT P
140 IF P = 0.1 THEN A = 1.303:GOTO 210
150 IF P = 0.25 THEN A = 1.236:GOTO 210
160 IF P = 0.5 THEN A = 1.158:GOTO 210
170 IF P = 1.0 THEN A = 1.046:GOTO 210
180 IF P = 2.0 THEN A = 0.8862:GOTO 210
190 IF P = 3.0 THEN A = 0.7666:GOTO 210
200 GOTO 110
210 PRINT "DO YOU WISH TO SPECIFY NATURAL FREQUENCY (F0)"
220 PRINT "OR PEAK FREQUENCY (FP)?"
230 INPUT FR$
240 PRINT "ENTER VALUE OF THIS FREQUENCY (HZ)"
250 INPUT F
260 IF FR$ = "F0" THEN F0=F:FP=F0*SQR(1-(A*A/2)):GOTO 310
270 IF FR$ = "FP" THEN F0=F*SQR(2/(2-A*A)):FP=F:GOTO 310
280 PRINT "ERROR" :GOTO 210
290 PRINT "REQUIRED NATURAL FREQUENCY (HZ)?"
```

```
300 INPUT F0
310 W0 = 2*π*F0
320 PRINT "CHOOSE A VALUE FOR C1 (MICROFARADS)"
330 INPUT C1
340 C2 = (A*A*C1)/4
350 R = 1000/(W0*SQR(C1*C2))
360 PRINT "A="A
370 PRINT "FP="FP"HZ"
380 PRINT "F0="F0"HZ"
390 PRINT "C1="C1"MICROFARADS"
400 PRINT "C2="C2"MICROFARADS"
410 PRINT "R="R"KILOHMS"
420 END
```

*Specimen run* (*Bessel*)

```
WHICH TYPE OF FILTER IS REQUIRED?
BESSEL(BE),BUTTERWORTH(BU) OR CHEBYSHEV(CH)
?BE
REQUIRED NATURAL FREQUENCY (HZ)?
? 1000
CHOOSE A VALUE FOR C1 (MICROFARADS)
? .1
A= 1.732
FP= 0 HZ
F0= 1000 HZ
C1= .1 MICROFARADS
C2= .0749956 MICROFARADS
R= 1.8378169 KILOHMS
```

*Specimen run* (*Butterworth*)

```
WHICH TYPE OF FILTER IS REQUIRED?
BESSEL(BE),BUTTERWORTH(BU) OR CHEBYSHEV(CH)
?BW
WHICH TYPE OF FILTER IS REQUIRED?
BESSEL(BE),BUTTERWORTH(BU) OR CHEBYSHEV(CH)
?BU
REQUIRED NATURAL FREQUENCY (HZ)?
? 400
CHOOSE A VALUE FOR C1 (MICROFARADS)
? .047
A= 1.414
FP= 0 HZ
F0= 400 HZ
C1= .047 MICROFARADS
C2= .023492903 MICROFARADS
R= 11.9740997 KILOHMS
```

of filter is required (lines 40 to 60). If either Bessel or Butterworth is specified the value of $\alpha$ (A in the program) is immediately defined (lines 70 and 80). For the Chebyshev filter a choice of peak gains corresponding to Table 6.1 is offered and the appropriate value of $\alpha$ is selected (lines 110 to 190). Non-allowed values cause a return to line 110.

## Second order active filters 69

*Specimen* (*Chebyshev*)

```
WHICH TYPE OF FILTER IS REQUIRED?
BESSEL(BE),BUTTERWORTH(BU) OR CHEBYSHEV(CH)
?CH
WHAT IS THE REQUIRED PEAK GAIN?
0.1,0.25,0.5,1,2 OR 3 DB?
? 2
DO YOU WISH TO SPECIFY NATURAL FREQUENCY (F0)
OR PEAK FREQUENCY (FP)?
?FP
ENTER VALUE OF THIS FREQUENCY (HZ)
? 12000
CHOOSE A VALUE FOR C1 (MICROFARADS)
? 1E-03
A= .8862
FP= 12000 HZ
F0= 15398.228 HZ
C1= 1E-03 MICROFARADS
C2= 1.9633761E-04 MICROFARADS
R= 23.3263956 KILOHMS

WHICH TYPE OF FILTER IS REQUIRED?
BESSEL(BE),BUTTERWORTH(BU) OR CHEBYSHEV(CH)
?CH
WHAT IS THE REQUIRED PEAK GAIN?
0.1,0.25,0.5,1,2 OR 3 DB?
? 2
DO YOU WISH TO SPECIFY NATURAL FREQUENCY (F0)
OR PEAK FREQUENCY (FP)?
?FO
ENTER VALUE OF THIS FREQUENCY (HZ)
? 12000
ERROR
DO YOU WISH TO SPECIFY NATURAL FREQUENCY (F0)
OR PEAK FREQUENCY (FP)?
?F0
ENTER VALUE OF THIS FREQUENCY (HZ)
? 12000
CHOOSE A VALUE FOR C1 (MICROFARADS)
? 1E-03
A= .8862
FP= 9351.72542 HZ
F0= 12000 HZ
C1= 1E-03 MICROFARADS
C2= 1.9633761E-04 MICROFARADS
R= 29.9320964 KILOHMS
```

For Chebyshev filters the operator may well wish to specify the frequency of the peak rather than the natural frequency. A choice is provided in lines 210 to 230. The one which is *not* specified is calculated using Equation (6.12) in line 260 or 270 as appropriate. For Bessel and Butterworth filters only the natural frequency can be entered (lines 290 and 300). The natural frequency is converted to angular form in line 310 and, after requesting a suitable value for $C_1$

(lines 320 and 330), $C_2$ and $R$ are calculated using Equations (6.13) and (6.14) (lines 350 and 340) and the parameters displayed.

Demonstration runs are provided for each type of filter. Notice that incorrect use of BW for Butterworth was detected as was FO as opposed to F0 in the second Chebyshev run. These two runs use the same nominal frequency value but in the first case it is interpreted as a peak value and as a natural frequency in the second.

The zero value for the frequency of peak response in the case of Bessel and Butterworth filters confirms the absence of a peak in the responses of these characteristics. (The peak is effectively at zero frequency.)

*(2) Equal component filter*
Although Equations (6.13) and (6.14) were easy to apply, difficulties can arise in view of the *ratio* of two capacitances which determines $\alpha$. A convenient value for $C_1$ may lead to an awkward one for $C_2$, and for low damping coefficients the required ratio can become large.

An alternative is to assign the same values to the two resistors and also to the two capacitors, that is

$$R_1 = R_2 = R \quad \text{and} \quad C_1 = C_2 = C$$

The value of $A_o$ is no longer restricted to unity, of course.

From Equations (6.10) and (6.11)

$$\omega_o = 1/(RC) \tag{6.15}$$

and

$$\alpha = [RC + RC + (1 - A_o)RC]/(RC)$$
$$= 3 - A_o \tag{6.16}$$

The natural frequency is determined by the time constant of $R$ and $C$ and the damping coefficient is determined only by $A_o$ and adjustment is possible without affecting the other parameters of the filter ($\omega_o$ will be unchanged but there will be a small effect on $\omega_p$).

It is unlikely that the value of $A_o$ required by Equation (6.16) will coincide with the overall gain required. Any required correction can be made by means of an additional, non frequency dependent, amplifying stage.

Program P(6.3) is very similar to P(6.2), in fact the same line numbers have been used where appropriate, but Equations (6.15) and (6.16) have been substituted (lines 350 and 355) to cater for the equal component value case. In view of the similarity only a single specimen run is presented.

```
10 REM PROGRAM P(6.3)
20 REM SALLEN KEY EQUAL COMPONENT FILTER DESIGN
30 PRINT ""
40 PRINT "WHICH TYPE OF FILTER IS REQUIRED?"
50 PRINT "BESSEL(BE),BUTTERWORTH(BU) OR CHEBYSHEV(CH)"
60 INPUT F$
70 IF F$ = "BE" THEN A = 1.732:GOTO 290
80 IF F$ = "BU" THEN A = 1.414:GOTO 290
90 IF F$ = "CH" THEN 110
100 GOTO 40
110 PRINT "WHAT IS THE REQUIRED PEAK GAIN?"
120 PRINT "0.1,0.25,0.5,1,2 OR 3 DB?"
130 INPUT P
140 IF P = 0.1 THEN A = 1.303:GOTO 210
150 IF P = 0.25 THEN A = 1.236:GOTO 210
160 IF P = 0.5 THEN A = 1.158:GOTO 210
170 IF P = 1.0 THEN A = 1.046:GOTO 210
180 IF P = 2.0 THEN A = 0.8862:GOTO 210
190 IF P = 3.0 THEN A = 0.7666:GOTO 210
200 GOTO 110
210 PRINT "DO YOU WISH TO SPECIFY NATURAL FREQUENCY (F0)"
220 PRINT "OR PEAK FREQUENCY (FP)?"
230 INPUT FR$
240 PRINT "ENTER VALUE OF THIS FREQUENCY (HZ)"
250 INPUT F
260 IF FR$ = "F0" THEN F0=F:FP=F0*SQR(1-(A*A/2)):GOTO 310
270 IF FR$ = "FP" THEN F0=F*SQR(2/(2-A*A)):FP=F:GOTO 310
280 PRINT "ERROR" :GOTO 210
290 PRINT "REQUIRED NATURAL FREQUENCY (HZ)?"
300 INPUT F0
310 W0 = 2*π*F0
320 PRINT "CHOOSE A VALUE FOR C (MICROFARADS)"
330 INPUT C
350 R = 1000/(W0*C)
355 A0 = 3-A
360 PRINT "A="A
365 PRINT "A0="A0
370 PRINT "FP="FP"HZ"
380 PRINT "F0="F0"HZ"
390 PRINT "C="C"MICROFARADS"
410 PRINT "R="R"KILOHMS"
420 END
```

## Specimen run

```
WHICH TYPE OF FILTER IS REQUIRED?
BESSEL(BE),BUTTERWORTH(BU) OR CHEBYSHEV(CH)
?BU
REQUIRED NATURAL FREQUENCY (HZ)?
? 400
CHOOSE A VALUE FOR C (MICROFARADS)
? .047
A= 1.414
A0= 1.586
FP= 0 HZ
F0= 400 HZ
C= .047 MICROFARADS
R= 8.46568847 KILOHMS
```

## 6.3.2 High pass filter

A second order high pass filter can be realized by making $Z_1$ and $Z_2$ of Figure 6.4 capacitive ($C_1$ and $C_2$ respectively) and $Z_3$ and $Z_4$ resistive ($R_1$ and $R_2$ respectively). Equation (6.7) becomes:

$$\frac{V_o}{V_i}(s) = \frac{A_o R_1 R_2}{1/(sC_1 sC_2) + R_1/(sC_2) + R_1 R_2 + R_1/(sC_1) + R_2(1 - A_o)/(sC_1)}$$

$$= \frac{A_o s^2}{s^2 + s[1/(C_2 R_2) + 1/(C_1 R_2) + (1 - A_o)/(C_1 R_1)] + 1/(R_1 C_1 R_2 C_2)}$$

(6.17)

which is of the general form:

$$\frac{V_o}{V_i}(s) = \frac{A_o s^2}{s^2 + \alpha \omega_o s + \omega_o^2} \qquad (6.18)$$

where

$$\omega_o = 1/\sqrt{(R_1 C_1 R_2 C_2)} \qquad (6.19)$$

as before, and

$$\alpha = \frac{R_1 C_1 + R_1 C_2 + (1 - A_o)C_2 R_2}{\sqrt{(R_1 C_1 R_2 C_2)}} \qquad (6.20)$$

which is slightly different from the low pass case (compare Equation (6.11)).

The $s^2$ ($= -\omega^2$ when $j\omega$ is substituted for $s$) term in the denominator of Equations (6.17) and (6.18) ensures that the gain increases with frequency. In fact the response can be shown to be a mirror image of that presented in Figure 6.5, tending to a constant gain at high frequencies. It is important to realize that for all practical operational amplifiers this 'constant' gain will be subject to high frequency roll-off and slew rate limiting (Sections 3.2 and 3.3).

The frequency of the peak of the response for the lightly damped cases will be higher than $\omega_o$ in this case; the 'correction factor' is the reciprocal of that given by Equation (6.12).

The phase response is similar but displaced by 180°. That is, there is a phase lead which tends to 180° at very low frequencies, a lead of

90° at $\omega_o$ and zero phase shift at very high frequencies (ignoring any lag imposed by the operational amplifier itself).

For the unity gain case, with $R_1 = R_2 = R$ we have

$$\omega_o = 1/[R\sqrt{(C_1 C_2)}] \qquad (6.21)$$

as before, and

$$\alpha = (C_1 + C_2)/\sqrt{(C_1 C_2)} \qquad (6.22)$$

For the equal component filter $\omega_o$ and $\alpha$ are given, as before, by Equations (6.15) and (6.16).

### 6.3.3 Band pass filters

A typical band pass filter response is shown in Figure 6.6. The frequency of the peak of response (the centre frequency) is $f_o$ and the higher and lower 3 dB points are $f_h$ and $f_l$ respectively. Notice that a logarithmic scale is used for frequency and $f_o$ is the *geometric* mean of $f_l$ and $f_h$, that is:

$$f_o = \sqrt{(f_l f_h)}$$

The selectivity of the filter, or 'sharpness' of the response, is often specified by the $Q$ ('quality factor') of the circuit. This is defined by the ratio of the centre frequency and the bandwidth, or

$$Q = f_o/(f_h - f_l) \qquad (6.23)$$

A simple band pass filter can be obtained from the general model of Figure 6.4. $Z_1$ and $Z_3$ must be made resistive ($Z_1 = R_1, Z_3 = R_2$) and $Z_2$ must be capacitive ($Z_2 = 1/(sC_1)$). For $Z_4$ a parallel resistance/capacitance combination is required.

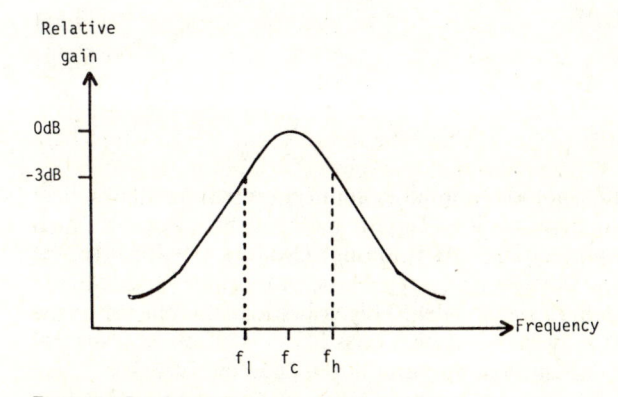

Figure 6.6 Band pass filter response

Let these be $R_3$ and $C_2$; then

$$Z_4 = \frac{R_3/(sC_2)}{R_3 + 1/(sC_2)} = \frac{R_3}{1 + sC_2R_3}$$

Substituting these values in Equation (6.7) and rearranging gives:

$$\frac{V_o}{V_i}(s) = \frac{A_o s/(C_2 R_1)}{s^2 + s[1/(C_1R_1) + 1/(C_1R_2) + 1/(C_2R_1) + 1/(C_2R_3) + (1-A_o)/(C_2R_2)] + 1/(C_1C_2R_3)(1/R_1 + 1/R_2)} \quad (6.24)$$

This unwieldy equation can be simplified by letting

$$R_1 = R_2 = R \quad \text{and} \quad C_1 = C_2 = C$$

Hence:

$$\frac{V_o}{V_i}(s) = \frac{A_o s/(CR)}{s^2 + s(5 - A_o)/(CR) + 2/(CR)^2}. \quad (6.25)$$

This is of the general form

$$\frac{V_o}{V_i}(s) = \frac{A'_o \alpha \omega_o s}{s^2 + \alpha \omega_o s + \omega_o^2} \quad (6.26)$$

which is the response of a second order band pass filter. Notice that $A'_o$, which is the gain at the centre frequency, is not equal to $A_o$ (as was the case for the low and high frequency gains of the low and high pass filters respectively).

The damping coefficient $\alpha$ can be shown to be the reciprocal of the $Q$ of the filter. Therefore, from Equations (6.25) and (6.26) we have

$$\omega_o = \sqrt{2}/(CR) = 2\pi f_o \quad (6.27)$$

$$\alpha = 1/Q = (5 - A_o)/\sqrt{2} \quad (6.28)$$

Also

$$A'_o = A_o/(5 - A_o) \quad (6.29)$$

Several points should be noted in connection with this filter. The low and high frequency roll-off (Figure 6.6) tends to a slope of 20 dB per decade. However, particularly for high $Q$ values, the slopes can be much greater in the vicinity of $f_o$. Notice this value of 20 dB per decade for a second order filter. This is because the characteristic 40 dB per decade of low and high pass filters is effectively 'shared' between the rising and falling parts of the characteristic.

Equation (6.28) imposes practical limits on the range of $Q$ values

Second order active filters 75

which can be obtained. For high values of $Q$ ($A_o$ approaching 5) the filter performance becomes extremely sensitive to parameter variations, particularly $A_o$ itself. For this reason $Q$ values greater than 10, or at the most 20, are not recommended. For higher $Q$ values several cascaded stages (Ref. 6.8), a negative emittance circuit (Ref. 6.9) or a state variable filter (Section 6.4) should be used.

Similarly, very low values of $Q$ lead to small and even negative values of $A_o$. In this (wide bandwidth) case, separate low and high pass filters, as discussed in the previous sections, are to be preferred.

Program P(6.4) uses Equations (6.27) to (6.29) for the design of a simple band pass filter. Either $Q$ or 3 dB bandwidth may be specified; the choice is made in lines 60 to 90 and a conversion to $Q$, if required, is made in line 120. For excessively low and high $Q$ values a message is printed and the program is terminated (lines 500 and 510). For $Q$ values between 10 and 20 a warning on sensitivity to parameter values is printed (lines 520 and 525). Circuit parameters are calculated and displayed (lines 528 to 630) before the operator chooses a suitable capacitance value (lines 700 and 710). Finally, the required resistance value is displayed.

The specimen runs show some typical cases. Notice that the centre frequency gain $A'_o$ (A1 in the program) becomes large for high $Q$

```
10 REM PROGRAM P(6.4)
20 REM BANDPASS FILTER DESIGN
30 PRINT "⏎"
40 PRINT "CENTRE FREQUENCY (HZ)"
50 INPUT F0
60 PRINT "SPECIFY Q OR 3DB BANDWIDTH (BW)?"
70 INPUT S$
80 IF S$ = "Q" THEN INPUT "Q";Q:GOTO 500
90 IF S$ <> "BW" THEN 60
100 PRINT"ENTER REQUIRED BANDWIDTH (HZ)"
110 INPUT BW
120 Q = F0/BW
500 IF Q<1 THEN PRINT "NOT SUITABLE FOR Q=";Q:END
510 IF Q>20 THEN PRINT "NOT SUITABLE FOR Q=";Q:END
515 IF Q < 10 THEN 530
520 PRINT"THIS CIRCUIT IS SENSITIVE TO PARAMETER"
525 PRINT "VARIATIONS FOR HIGH Q VALUES"
528 BW = F0/Q
530 PRINT "F0=";F0
540 PRINT "Q=";Q
550 PRINT "BW=";BW
600 A0 = 5 - (1.414/Q)
610 PRINT "A0=";A0
620 A1 = A0/(5 - A0)
630 PRINT "A1=";A1
700 PRINT "CHOOSE A CAPACITANCE VALUE (MICROFARADS)"
710 INPUT C
720 R = 1000/(F0*1.414*C)
730 PRINT "R=";R"KILOHMS"
```

## Specimen run

```
CENTRE FREQUENCY (HZ)
? 400
SPECIFY Q OR 3DB BANDWIDTH (BW)?
?Q
Q? 10
THIS CIRCUIT IS SENSITIVE TO PARAMETER
VARIATIONS FOR HIGH Q VALUES
F0= 400
Q= 10
BW= 40
A0= 4.85857865
A1= 34.3553393
CHOOSE A CAPACITANCE VALUE (MICROFARADS)
? .1
R= 5.62697698 KILOHMS

CENTRE FREQUENCY (HZ)
? 400
SPECIFY Q OR 3DB BANDWIDTH (BW)?
?BW
ENTER REQUIRED BANDWIDTH (HZ)
? 20
THIS CIRCUIT IS SENSITIVE TO PARAMETER
VARIATIONS FOR HIGH Q VALUES
F0= 400
Q= 20
BW= 20
A0= 4.92928933
A1= 69.7106786
CHOOSE A CAPACITANCE VALUE (MICROFARADS)
? .1
R= 5.62697698 KILOHMS

CENTRE FREQUENCY (HZ)
? 5000
SPECIFY Q OR 3DB BANDWIDTH (BW)?
?BW
ENTER REQUIRED BANDWIDTH (HZ)
? 7500
NOT SUITABLE FOR Q= .66666667

CENTRE FREQUENCY (HZ)
? 5000
SPECIFY Q OR 3DB BANDWIDTH (BW)?
?BW
ENTER REQUIRED BANDWIDTH (HZ)
? 5000
F0= 5000
Q= 1
BW= 5000
A0= 3.58578644
A1= 2.53553391
CHOOSE A CAPACITANCE VALUE (MICROFARADS)
? .02
R= 2.25079079 KILOHMS
```

values. This means, inevitably, that the amplifier will become saturated by relatively small input signals at this frequency. For wide bandwidths, however, the centre frequency gain is actually less than the basic voltage follower gain $A_o$ (see the last specimen run).

## 6.4 State variable filters

Originally, the principal application of operational amplifiers was in analogue computers. Configurations of operational integrators and summers enabled rapid continuous solutions to be obtained for a wide range of complex differential equations (see, for example, Reference 6.10). Their use has now been largely superseded by the use of digital computer simulation techniques.

However, one application of analogue computing techniques is still widely used in the form of the state variable filter. These can be of any order, but here attention will be devoted to the second order version (Figure 6.7). This circuit bears a strong resemblance to an analogue simulation diagram for a linear second order system such as a feedback controller. (Hence the title; the outputs correspond to the state variables of the system—see, for example, Reference 6.4, p. 312.)

The first stage is a differential amplifier which is complicated slightly by the use of the non-inverting input to add $V_i$ and $V_{OB}$. In Figure 6.7, several groups of components have been given equal values. This is not essential but helps to simplify the analysis. Since it is assumed that no current flows into the amplifier, the voltage at the non-inverting input can be determined, by potentiometer action, as

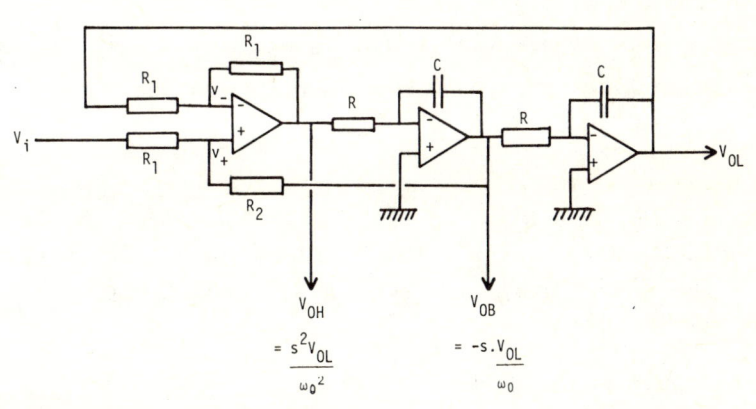

$$v_+ = V_i R_2/(R_1 + R_2) + V_{OB} R_1/(R_1 + R_2)$$

*Figure 6.7* Second order state variable filter

78  Introduction to active filters

Noting that, for a high open loop gain, $v_+ \simeq v_-$ and substituting in the equations of Section 2.4,

$$V_{OH} = \frac{2}{(R_1 + R_2)}(R_2 V_i + R_1 V_{OB}) - V_{OL} \qquad (6.30)$$

The second and third stages are simple integrators (Section 5.2). In Laplace form, Equation (5.1) may be written

$$V_o(s) = -\omega_o/s$$

where $\omega_o = 1/(RC)$ and the initial condition has been neglected. Hence, in Figure 6.7:

$$V_{OB} = -s\frac{V_{OL}}{\omega_o} \quad \text{and} \quad V_{OH} = s^2 \frac{V_{OL}}{\omega_o^2} \qquad (6.31\text{a,b})$$

Substituting so that (6.30) provides $V_{OL}$ as a function of $V_i$:

$$\frac{V_{OL}}{V_i}(s) = \frac{2[R_2/(R_1 + R_2)]\omega_o^2}{s^2 + 2[R_1/(R_1 + R_2)]\omega_o s + \omega_o^2} \qquad (6.32)$$

which is of exactly the same form as Equation (6.9), where

$$A_o = 2R_2/(R_1 + R_2)$$
$$\alpha = 2R_1/(R_1 + R_2)$$

and $\omega_o = 1/(RC)$ as for the integrators.

That is, the relation between $V_i$ and $V_{OL}$ is precisely that of a second order low pass filter (hence the notation).

The effective output at $V_{OB}$ may be obtained from Equation (6.32) by means of (6.31a) as:

$$\frac{V_{OB}(s)}{V_i} = -\frac{2[R_2/(R_1 + R_2)]\omega_o s}{s^2 + 2[R_1/(R_1 + R_2)]\omega_o s + \omega_o^2} \qquad (6.33)$$

which is of the same form as Equation (6.26) with

$$A'_o = R_2/R_1$$

and

$$\alpha = 2R_1/(R_1 + R_2)$$

as before.

Therefore output $V_{OB}$ provides a band pass response relative to $V_i$. Similarly,

$$V_{OH}(s) = \frac{s^2}{\omega_o^2} V_{OL}(s)$$

$$= \frac{2[R_2/(R_1+R_2)]s^2}{s^2 + 2[R_1/(R_1+R_2)]\omega_o s + \omega_o^2} \qquad (6.34)$$

which is of the same form as Equation (6.18) with

$$A_o = 2R_2/(R_1+R_2)$$

and

$$\alpha = 2R_1/(R_1+R_2)$$

as before, and a high pass response.

Although a second order state variable filter requires three amplifiers, it has some advantages. First, it is much less sensitive to parameter variations than the single amplifier configuration. Band pass filters with a $Q$ of 50 or more can readily be realized. Secondly, the provision of low, high and band pass functions in a single building block is particularly convenient from a practical point of view.

## 6.5 Band rejection filter

The final class of filter to be discussed is that which rejects, or at least substantially reduces, signals at a particular frequency or within a specified range of frequencies.

One approach (see Reference 6.11 for further details) is to subtract the output of a suitable band pass filter from the required signal, as shown in Figure 6.8. Since the band pass filter negates $V_i$, the subtraction is effectively performed by an operational adder. The result is a minimum in the output $V_o$ for those frequencies that provide the maximum output from the band pass filter.

From Figure 6.8 and Equation (6.26), and taking $A_o' = 1$ for the band pass filter, it follows that:

$$\frac{V_o(s)}{V_i} = -\frac{A_o(s^2 + \omega_o^2)}{s^2 + \alpha\omega_o s + \omega_o^2} \qquad (6.35)$$

which is the standard form of a band rejection characteristic.

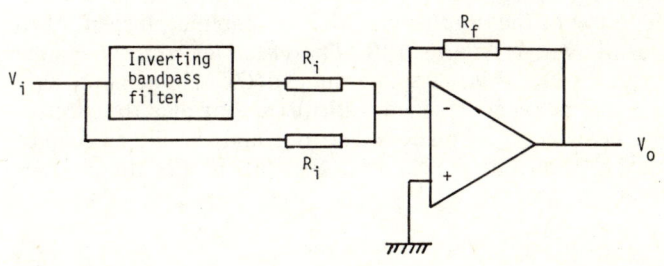

*Figure 6.8* Use of a band pass filter to realize a band rejection function

80  Introduction to active filters

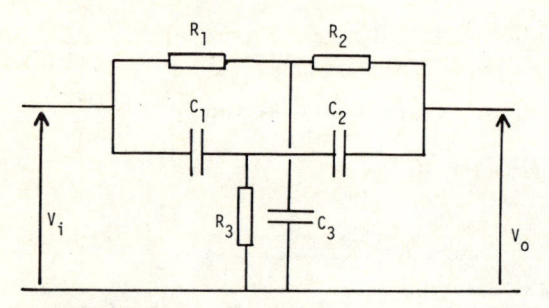

*Figure 6.9* Basic twin-T network

The twin-T network, as shown in passive form in Figure 6.9, is often used to realize a 'notch' characteristic (a narrow bandwidth filter intended for the rejection of one specific frequency). Since there are three capacitors, the transfer function of this network in its general form is of third order. However, provided $R_1C_1 = R_2C_2$ (which normally applies to the component values chosen for a practical realization) it can be shown (see Reference 6.12) that cancellation occurs and the transfer function is reduced to a second order one.

In particular, it is customary to make

$$R_1 = R, \quad R_2 = kR \quad \text{and} \quad R_3 = [k/(1+k)]R$$
$$C_1 = C, \quad C_2 = C/k \quad \text{and} \quad C_3 = C(1 + 1/k) \tag{6.36}$$

since this greatly simplifies the analysis. It is sometimes called the 'potentially symmetric' case since making $k = 1$ causes the network to become symmetrical. It can be shown (Reference 6.12 again) that in this case $\alpha$ (Equation (6.35)) becomes $2(1 + 1/k)$ which means that the maximum value of $Q$ ($= 1/\alpha$) is obtained as $k$ becomes very small and cannot in any case exceed 0.5.

A dramatic increase in $Q$ was obtained following a suggestion by Farrer (Ref. 6.13). This involved connecting the 'bottom' of the T to some fraction $m$ of the output, via a buffer amplifier, instead of to ground, as shown in Figure 6.10. The effective value of $\alpha$ now becomes $2(1 + 1/k)(1 - m)$ and large values of $Q$ can be achieved as $m$ approaches 1. Since $m$ now provides adequate control of the effective $Q$, $k$ can conveniently be made equal to 1 and the filter becomes symmetrical as shown by the values in Figure 6.10. For these values we have:

$$Q = \frac{1}{4(1-m)} \tag{6.37a}$$

Band rejection filter 81

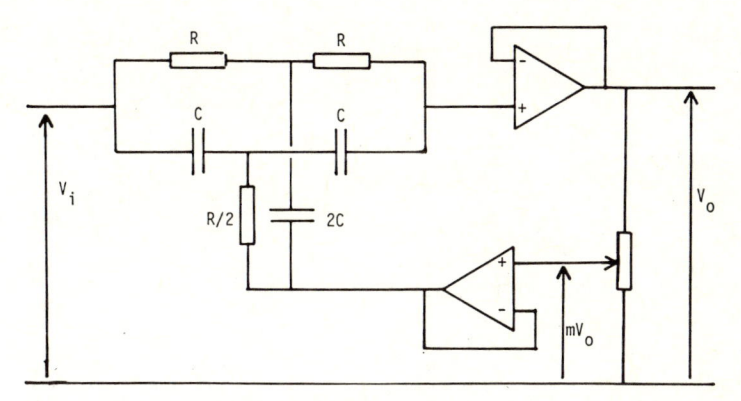

*Figure 6.10* Active version of twin-T notch filter

$$f_o = \frac{\omega_o}{2\pi} = \frac{1}{2\pi RC} \qquad (6.37b)$$

As *m* tends to zero the characteristic of the filter approaches that of the passive version (with a buffered output) and the *Q* tends to its minimum value of 0.25. Larger values of *m* allow a *Q* value of around 50 to be readily obtained.

Program P(6.5) enables suitable component values to be selected. As with program P(6.4) either *Q* or bandwidth may be entered and

```
10 REM PROGRAM P(6.5)
20 REM NOTCH FILTER DESIGN
30 PRINT ""
40 PRINT "CENTRE FREQUENCY (HZ)"
50 INPUT F0
60 PRINT "SPECIFY Q OR 3DB BANDWIDTH (BW)"
70 INPUT S$
80 IF S$ = "Q" THEN INPUT "Q";Q:GOTO 500
90 IF S$ <> "BW" THEN 60
100 PRINT"ENTER REQUIRED BANDWIDTH (HZ)"
110 INPUT BW
120 Q = F0/BW
500 IF Q<0.25 THEN PRINT "NOT SUITABLE FOR Q="Q:END
510 IF Q>50 THEN PRINT "NOT SUITABLE FOR Q="Q:END
520 BW = F0/Q
530 PRINT "F0="F0"HZ"
540 PRINT "Q="Q
550 PRINT "BW="BW"HZ
700 PRINT "CHOOSE A CAPACITANCE VALUE (MICROFARADS)"
710 INPUT C
720 R = 1000/(2*π*F0*C)
730 PRINT "R="R"KILOHMS"
740 M = 1 - 1/(4*Q)
750 PRINT "M="M
760 END
```

## Specimen run

```
CENTRE FREQUENCY (HZ)
? 400
SPECIFY Q OR 3DB BANDWIDTH (BW)
?Q
Q? 100
NOT SUITABLE FOR Q= 100


CENTRE FREQUENCY (HZ)
? 400
SPECIFY Q OR 3DB BANDWIDTH (BW)
?Q
Q? 50
F0= 400 HZ
Q= 50
BW= 8 HZ
CHOOSE A CAPACITANCE VALUE (MICROFARADS)
? .02
R= 19.8943679 KILOHMS
M= .995
```

warnings are printed for inappropriate values. A capacitance value is selected by the user (lines 700 and 710) and the required values of $R$ and $m$ are calculated from Equations (6.37a) and (6.37b) (lines 720 and 740). The specimen runs should be self-explanatory.

### 6.6 Exercises

**(6.1)** Using any graphics facilities which may be available on your computer, write programs to plot the gain/frequency and phase/frequency relationships specified by Equations (6.4) and (6.5). Use a logarithmic (dB) scale for gain, a linear scale for phase angle and a logarithmic scale for frequency in both cases. Compare your results with Figure 3.3.

**(6.2)** Extend the program specified in the previous question to cover high pass and band pass filters.

**(6.3)** Write a program that computes the modulus of Equation (6.9), substituting $\alpha$ for $2\zeta$. Use your program to verify and extend the required values of $\alpha$ for a given peak in the gain/frequency response.

**(6.4)** Use program P(2.2) to extend programs P(6.3) and (6.4) so that suitable values for $R_f$ and $R_i$ are printed.

**(6.5)** Extend programs P(6.2) and P(6.3) to cover high pass filters for which the relevant equations are given in sub-section 6.3.2.

**(6.6)** Using the formulae derived in Section 6.4, write programs on the lines of P(6.2), P(6.3) and P(6.4) for the design of state variable filters.

## 6.7 References

6.1 BOWRON, P. and STEPHENSON, F. W. *Active filters for communications and instrumentation.* McGraw-Hill 1979, p. 251
6.2 JACOB, J. M. *Applications and Design with Analog Integrated Circuits.* Reston Publishing 1982, pp 331–4
6.3 SAVANT, C. J. *Fundamentals of the Laplace Transformation.* McGraw-Hill 1962
6.4 DORF, R. C. *Modern Control Systems.* Addison Wesley 1980
6.5 BROWN, J. and GLAZIER, E. V. D. *Telecommunications.* Chapman and Hall 1974, pp 80–3
6.6 BARBE, E. C. *Linear Control Systems.* International Textbook Co 1963, p. 274
6.7 SALLEN, R. P. and KEY, E. L. A practical method of designing R C active filters. *IRE Transactions on circuit theory*, Vol. CT2 No. 1, pp 74–85, March 1955
6.8 As 6.2, pp 373–6
6.9 CLAYTON, G. B. *Linear integrated circuit applications.* Macmillan 1975, pp 65–8
6.10 KORN, G. A. and KORN, T. M. *Electronic Analog and Hybrid Computers.* McGraw-Hill, New York, 1964
6.11 As 6.2, pp 376–81
6.12 MOSCHYTZ, G. S. A general approach to twin-T design and its application to hybrid integrated linear active networks. *Bell System Technical Journal* **49**, No. 6, pp 1105–49, July–August 1970
6.13 FARRER, W. A simple active filter with independent control over the pole and zero locations. *Electronic Engineering* **39**, No. 470, pp 219–22, April 1967

Chapter 7
# Non-linear circuits

## 7.1 Preliminary comments

With the exception of comparators, only linear applications of operational amplifiers have been discussed in the previous chapters. This means that output magnitude is directly proportional to input magnitude. In the case of multi-input circuits, such as summers, any change in output will still be proportional to any one input change, even in the case where inputs are added with different scaling factors.

In many applications, linear operation is an important requirement. Examples are high quality audio amplifiers and signal analysis systems. The negative feedback which is inherent in operational amplifier configurations ensures highly linear operation. However, there are applications where deliberately introduced non-linearity is required. These include rectification and amplitude limiting of signals, waveform shaping (e.g. from triangular to sinusoidal) and dynamic range compression and expansion for communication purposes.

Although a wide variety of non-linear circuits has been described, two categories can be distinguished:

(1) Those which are based on the use of diodes as switches. In simple cases, such as rectification, a single switching operation may be all that is required. Some examples will be discussed in Sections 7.2 and 7.3. More generally, any non-linear characteristic can be realized as a series of piece-wise linear approximations, using diodes to select each required segment of the characteristic. This approach will be discussed in Section 7.4.

(2) Those which exploit the inherent non-linear characteristic of a particular device. In particular, the bipolar transistor is widely used to generate logarithmic and antilogarithmic functions as discussed in Section 7.5.

## 7.2 Simple limiting

The transfer characteristic of an ideal limiter is shown in Figure 7.1. For input signals between $V_{Li+}$ and $V_{Li-}$, the output $V_o$ is directly

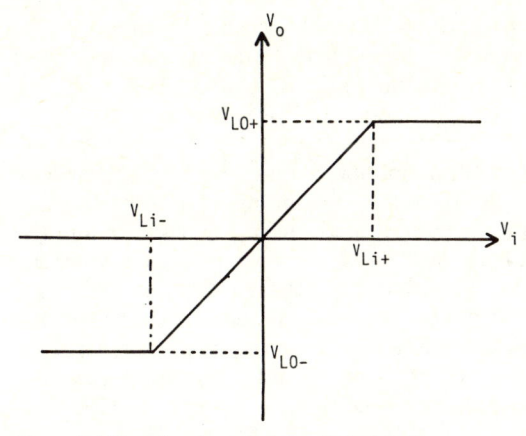

Figure 7.1 Transfer characteristic of ideal limiter

proportional to $V_i$ with a gain which may be selected in the usual way. For inputs which exceed these values the output remains constant at the appropriate value of $V_{Lo+}$ or $V_{Lo-}$. Circuits of this kind are particularly useful, for example in control systems, where it is necessary to protect later parts of the system from overload.

A variety of circuits which can realize, or approximate, this characteristic have been proposed. Two of the more important will now be discussed.

A simple feedback limiter is shown in Figure 7.2. For small values of $V_i$, and hence $V_o$, both diodes will be reverse biased and the circuit

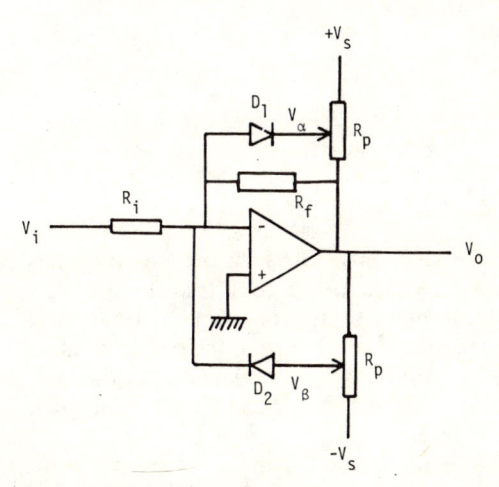

Figure 7.2 Simple feedback limiter

operates as a conventional operational amplifier with a gain of $-R_f/R_i$. The negative sign means that the characteristic of Figure 7.1 will be inverted. The potentiometers $R_p$ connected to $+V_s$ and $-V_s$ represent an additional load on the output of the amplifier which will not otherwise affect its performance. $\alpha R_p$ and $\beta R_p$ represents the portion of $R_p$ between the amplifier output and the appropriate diode. Adjustment of $\alpha$ and $\beta$ allows the limiting levels to be selected.

As $V_i$ becomes more positive, $V_o$ becomes more negative and the reverse bias $V_\alpha$ on the cathode of diode $D_1$ will be reduced. Since the anode of this diode is connected to the virtual earth point it will start to conduct when $V_\alpha$ becomes negative. When this happens, the effective feedback resistance is reduced to $\alpha R_p$ in parallel with $R_f$. Similarly, when $V_i$ becomes sufficiently negative for $D_2$ to conduct, the effective feedback resistance becomes $\beta R_p$ in parallel with $R_f$.

From Figure 7.2, before the diode conducts

$$V_\alpha = V_o + \alpha(V_s - V_o)$$

since current flow into a reverse biased diode is negligible.

Limiting occurs when $V_\alpha = 0$ or

$$V_o = -V_s[\alpha/(1-\alpha)] \tag{7.1}$$

Since the amplifier gain, before limiting occurs, is $-R_f/R_i$ this corresponds to

$$V_{Li+} = V_s \frac{R_i}{R_f}\left(\frac{\alpha}{1-\alpha}\right) \tag{7.2}$$

where $V_{Li+}$ is defined in Figure 7.1.

Similarly

$$V_{Li-} = -V_s \frac{R_i}{R_f}\left(\frac{\beta}{1-\beta}\right) \tag{7.3}$$

The simple feedback limiter, although useful, does not realize the ideal characteristic of Figure 7.1 for several reasons.

(1) Practical diodes require an appreciable forward bias (typically 0.5 volt) before significant conduction occurs (see Figure 7.3). Equation 7.1 is, therefore, in error by this amount. Normally this effect can be accommodated by a small change in the required values of $\alpha$ and $\beta$. However, the forward voltage of semiconductor diodes is temperature dependent and this will be reflected in temperature dependence of the limiting levels.

(2) Curvature of the diode characteristic (Figure 7.3) means that the discontinuities shown in Figure 7.1 will become rounded. In many

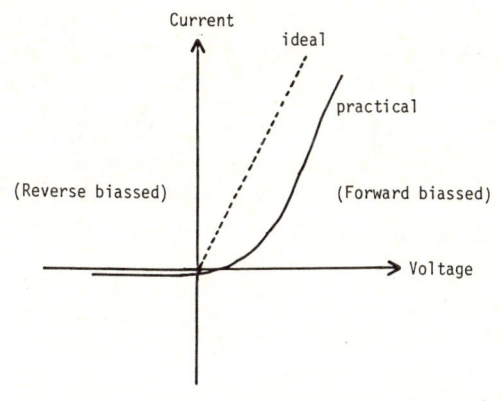

*Figure 7.3* Ideal and practical current/voltage characteristics for a diode with series resistance

applications this is acceptable and, in fact, represents a more accurate simulation of practical limiting effects.

(3) Since the effective feedback resistance, when limiting occurs, is not zero, the horizontal portions of the characteristic of Figure 7.1 will actually have a finite slope. However, by making $R_p$ small compared with $R_f$, the slope can be made acceptably small. For example, if $R_f = 1\,M\Omega$ and $R_p = 10\,k\Omega$, the maximum slope would be approximately 1 in 100 and significantly less for typical values of $\alpha$ and $\beta$ which are less than 0.5. For values of $\alpha$ and $\beta$ between 0.5 and 1 the diodes can never become forward biased.

In view of the effects described in (2) and (3) the resulting characteristic is sometimes described as 'soft' limiting. As before, this closely resembles limiting effects that occur in practice.

Program P(7.1) provides assistance in the design of feedback limiters. The first part of the program is concerned with determination of the minimum permissible feedback resistance and is based on the approach used in program P(2.1). The main difference is that allowance must be made for the additional loading of the two potentiometers (lines 90 to 120). The potentiometers are assumed to be of equal value and their parallel connection (with respect to the amplifier output) yields $R_p/2$. The minimum load calculation will be conservative since, for any limit value other than full scale, maximum output voltage will not be obtained.

After requesting the required circuit parameters a check is made to ensure that the proposed limits are appropriate (lines 230 and 270). The program calculates the required input resistance (line 200) and the potentiometer settings $\alpha$ and $\beta$ (A and B in the program) using

## Non-linear circuits

```
10 REM PROGRAM P(7.1)
20 REM FEEDBACK LIMITER DESIGN
30 REM CALCULATE MIN FEEDBACK RESISTANCE
40 PRINT ""
50 PRINT"SUPPLY VOLTAGE":INPUT VS
60 PRINT"MAX AMPLIFIER OUTPUT CURRENT (MA)":INPUT IM
70 PRINT"LOAD TO BE DRIVEN (KOHMS)"
80 INPUT L
90 PRINT "RESISTANCE OF POTENTIOMETERS  (KILOHMS)"
100 INPUT RP
110 REM CALCULATE EFFECTIVE LOAD RESISTANCE
120 RL = L*(RP/2)/(L+(RP/2))
130 IF IM*RL<=VS THEN PRINT"CAN'T DRIVE LOAD":GOTO 70
140 RM=(RL*VS)/(IM*RL-VS)
150 PRINT"MINIMUM RF IS "RM"KOHMS CHOOSE A VALUE"
160 INPUT RF
170 IF RF>=RM GOTO 190
180 PRINT "RF TOO SMALL":GOTO 150
190 PRINT"GAIN REQURED IN LINEAR REGION": INPUT G
200 RI =RF/G
210 PRINT "REQUIRED -VE LIMIT (AT AMPLIFIER OUTPUT)"
220 INPUT NL
230 IF NL<0 AND NL>-VS THEN 250
240 PRINT"OUT OF RANGE":GOTO 210
250 PRINT "REQUIRED +VE LIMIT (AT AMPLIFIER OUTPUT)"
260 INPUT PL
270 IF PL>0 AND PL<VS THEN 290
280 PRINT"OUT OF RANGE":GOTO 250
290 A = ABS(NL)/(VS+ABS(NL))
300 B = PL/(VS+PL)
310 PS = (RF*B*RP)/(RI*(RF+(B*RP)))
320 NS = (RF*A*RP)/(RI*(RF+(A*RP)))
330 PRINT"INPUT RESISTANCE ="RI"KOHMS"
340 PRINT"FEEDBACK RESISTANCE="RF"KOHMS"
350 PRINT"A="A
360 PRINT"B="B
370 PRINT"POSITIVE LIMIT SLOPE ="PS
380 PRINT"NEGATIVE LIMIT SLOPE ="NS
390 END
```

*Specimen runs*

```
SUPPLY VOLTAGE
? 15
MAX AMPLIFIER OUTPUT CURRENT (MA)
? 10
LOAD TO BE DRIVEN (KOHMS)
? 47
RESISTANCE OF POTENTIOMETERS   (KILOHMS)
? 10
MINIMUM RF IS  2.24522293 KOHMS CHOOSE A VALUE
? 10
GAIN REQURED IN LINEAR REGION
? 2
REQUIRED -VE LIMIT (AT AMPLIFIER OUTPUT)
? 5
OUT OF RANGE
```

```
REQUIRED -VE LIMIT (AT AMPLIFIER OUTPUT)
?-5
REQUIRED +VE LIMIT (AT AMPLIFIER OUTPUT)
? 7.5
INPUT RESISTANCE = 5 KOHMS
FEEDBACK RESISTANCE= 10 KOHMS
A= .25
B= .333333333
POSITIVE LIMIT SLOPE = .5
NEGATIVE LIMIT SLOPE = .4


SUPPLY VOLTAGE
? 15
MAX AMPLIFIER OUTPUT CURRENT (MA)
? 10
LOAD TO BE DRIVEN (KOHMS)
? 47
RESISTANCE OF POTENTIOMETERS   (KILOHMS)
? 1
CAN'T DRIVE LOAD
LOAD TO BE DRIVEN (KOHMS)
? 47
RESISTANCE OF POTENTIOMETERS   (KILOHMS)
? 10
MINIMUM RF IS  2.24522293 KOHMS CHOOSE A VALUE
? 100
GAIN REQURED IN LINEAR REGION
? 2
REQUIRED -VE LIMIT (AT AMPLIFIER OUTPUT)
?-5
REQUIRED +VE LIMIT (AT AMPLIFIER OUTPUT)
? 7.5
INPUT RESISTANCE = 50 KOHMS
FEEDBACK RESISTANCE= 100 KOHMS
A= .25
B= .333333333
POSITIVE LIMIT SLOPE = .0645161291
NEGATIVE LIMIT SLOPE = .0487804878
```

Equations (7.2) and (7.3). When limiting has occurred, the effective feedback resistance becomes $R_f$ in parallel with $\alpha R_p$ or $\beta R_p$. The corresponding slopes are calculated (lines 310 and 320) and displayed together with the other circuit parameters.

Two specimen runs are presented. In the first, potentiometer values equal to the feedback resistance were chosen, resulting in limit slopes which are far from negligible. The situation is improved in the second run by increasing the value of feedback resistance.

## 7.3 Precision limiting and rectification

For applications where the imperfections of the simple limiter, discussed in the previous section, are not acceptable, performance can be improved by including the switching diode (or diodes) within the amplifier feedback path, as shown in basic form in Figure 7.4.

The non-ideal characteristic of the diode (Figure 7.3) may be represented in a general form as a voltage drop $V_D$ which will be some function of the current flowing through it.

Analysis of Figure 7.4 in the usual way (Section 2.2) gives

$$\frac{V_i - v}{R_i} + \frac{V_o - v}{R_f} = 0$$

However,

$$V'_o = -Av$$

and

$$V_o = V'_o - V_D$$

Therefore

$$V_o = -Av - V_D$$

and

$$v = -(V_o + V_D)/A$$

Therefore, as $A$ becomes very large, $v$ tends to zero in the usual way and the error $V_D$ due to the diode is reduced by a factor of $A$.

This effect is exploited in the precision limiter shown in Figure 7.5. For small inputs, all diodes will be forward biased and the input of the second amplifier will closely follow the output of the first. Since the diode pairs $D_1$ and $D_2$ together with $D_3$ and $D_4$ are effectively back to back, errors due to their forward voltages will tend to cancel. Any residual error will be reduced by a factor of $A$ as explained above.

When the output of the first amplifier becomes sufficiently positive, $D_1$ will cease to conduct and, in view of the symmetry of the circuit, so

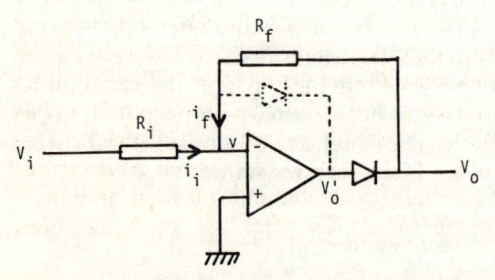

Figure 7.4 Incorporation of diode within the feedback path. The diode shown dotted protects the amplifier from saturation caused by positive inputs

Precision limiting and rectification 91

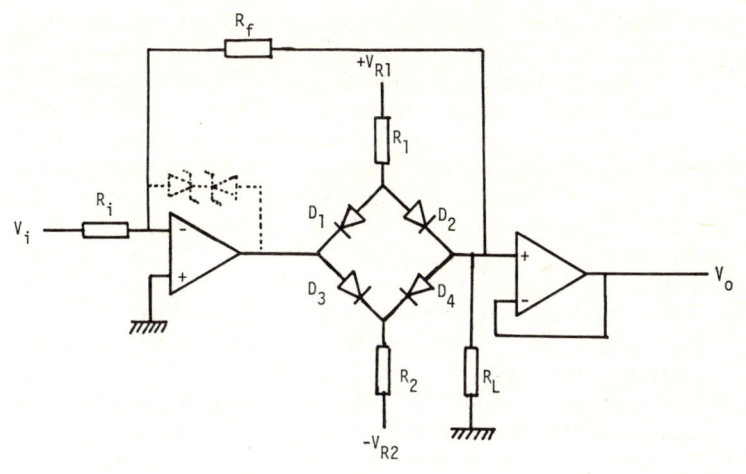

*Figure 7.5* Precision limiter circuit using a diode bridge (the Zener diodes shown dotted protect the amplifier from overload)

will $D_4$. The output $V_o$ is now determined by $+V_{R1}$ in association with $R_1$, $R_f$ and $R_L$.

Similarly, a negative output from the first amplifier eventually causes $D_3$ and $D_2$ to cease conducting. Analysis shows that the output saturation levels are given by

$$V_{LO+} = V_{R1} \frac{R_f R_L}{R_f R_L + R_f R_1 + R_L R_1} \tag{7.4}$$

$$V_{LO-} = V_{R2} \frac{R_f R_L}{R_f R_L + R_f R_2 + R_L R_2} \tag{7.5}$$

Notice that, when limiting has occurred, the output is independent of $V_i$ and a truly horizontal characteristic is obtained. (The effects of diode leakage current are negligible in view of the current flowing in the other two diodes which will be forward biased.) The limiting levels depend on $R_L$, whose effective value would be changed by the connection of an external load at this point. The buffer amplifier provides the isolation required to avoid this effect. When limiting has occurred, the first amplifier is effectively without feedback. This means that, for large values of $V_i$, it could be driven heavily into saturation. This will not normally cause damage but some amplifiers require an appreciable time to recover after the input has been reduced.

For high-speed circuits where this must be avoided the amplifier can be protected by a feedback limiter of the type discussed in Section

## 92 Non-linear circuits

7.2 or, more simply, a pair of Zener diodes as shown in Figure 7.5. The Zener voltages should be slightly greater than the required limiting values. For each extreme of $V_i$, one Zener diode will be forward biased (with a drop of around 0.5 volt) and the other will be in its reverse (Zener) region (with a voltage drop as specified by the manufacturer). For amplifier outputs that exceed the sum of these two voltages, the effective feedback resistance will become very small and further increase of output voltage will be prevented.

Program P(7.2) simply determines the required values of $R_1$ and $R_2$ (Figure 7.5) using rearrangements of Equations (7.4) and (7.5) (lines 150 and 160). A typical specimen run is presented.

It is often required to rectify alternating signals with good accuracy for measurement purposes. A common example is the provision of alternating current and voltage ranges on digital multimeters. The practical diode characteristic shown in Figure 7.3 would introduce serious errors, and values less than about 0.5 volt could not be measured at all! For this reason the 'precision rectifier circuit' is widely used.

The basic requirements are provided by the circuit of Figure 7.4 where, as already explained, the effects of non-ideal diode characteristics are reduced by the gain of the operational amplifier. This circuit provides precise half-wave rectification with the negative portions of the input providing a positive output, as shown in Figure 7.6. The positive parts of the input are lost and the amplifier can be protected from saturation caused by such inputs by means of the diode shown dotted in Figure 7.4. Negative outputs from the

```
10 REM PROGRAM P(7.2)
20 REM PRECISION LIMITER DESIGN
25 PRINT ""
30 PRINT "REFERENCE VOLTAGE"
40 INPUT VR
50 PRINT "FEEDBACK RESISTANCE (KILOHMS)"
60 INPUT RF
70 PRINT "LOAD RESISTANCE (KILOHMS)"
80 INPUT RL
90 PRINT "POSITIVE LIMIT VOLTAGE"
100 INPUT PL
110 PRINT "NEGATIVE LIMIT VOLTAGE"
120 INPUT NL
130 IF NL < 0 THEN NL = -NL
140 IF PL<=VR AND NL<=VR THEN 150
145 PRINT "LIMIT CANNOT EXCEED REFERENCE":GOTO 90
150 R1 = ((RF*RL)/(RF+RL))*((VR/PL)-1)
160 R2 = ((RF*RL)/(RF+RL))*((VR/NL)-1)
170 PRINT "R1="R1"KILOHMS"
180 PRINT "R2="R2"KILOHMS"
190 END
```

Precision limiting and rectification 93

## Specimen run

```
REFERENCE VOLTAGE
? 12
FEEDBACK RESISTANCE (KILOHMS)
? 27
LOAD RESISTANCE (KILOHMS)
? 10
POSITIVE LIMIT VOLTAGE
? 13
NEGATIVE LIMIT VOLTAGE
? 10
LIMIT CANNOT EXCEED REFERENCE
POSITIVE LIMIT VOLTAGE
? 8
NEGATIVE LIMIT VOLTAGE
? 6
R1= 3.64864865 KILOHMS
R2= 7.2972973 KILOHMS
```

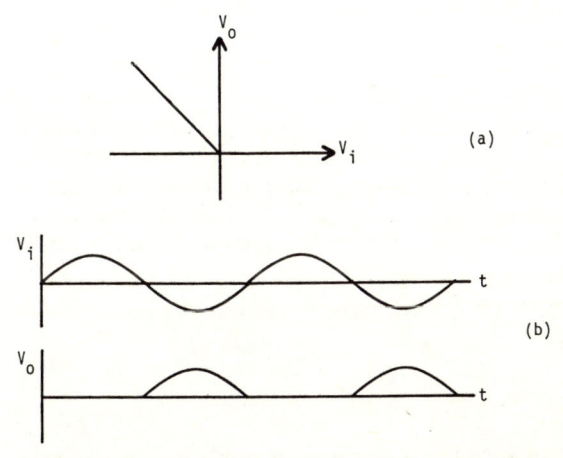

*Figure 7.6* Precision half-wave rectifier (a) Transfer characteristic (b) Effect on a sinusoidal input

amplifier forward bias the diode and cause a low effective feedback resistance.

Half-wave rectification is adequate for many applications but full-wave rectification (Figure 7.7) has the advantage of a higher ripple frequency and, therefore, less stringent filtering requirements. Full-wave rectification involves a relative sign change for one polarity of input only. The circuit of Figure 7.4 can readily be arranged to provide both positive and negative half cycles as shown in Figure 7.8. These can then be summed in a conventional differential amplifier (see Section 2.4).

94  Non-linear circuits

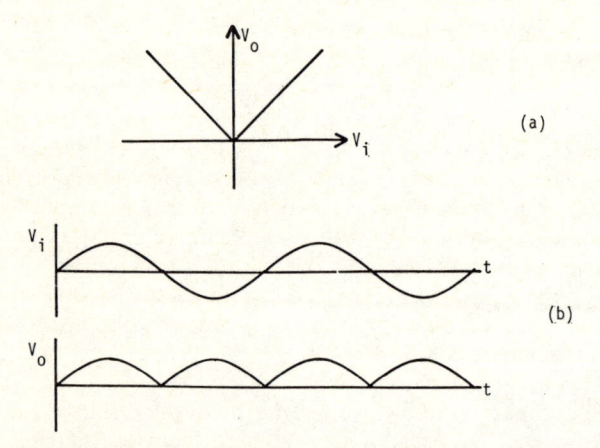

*Figure 7.7* Precision full-wave rectifier (a) Transfer characteristic (b) Effect on a sinusoidal input

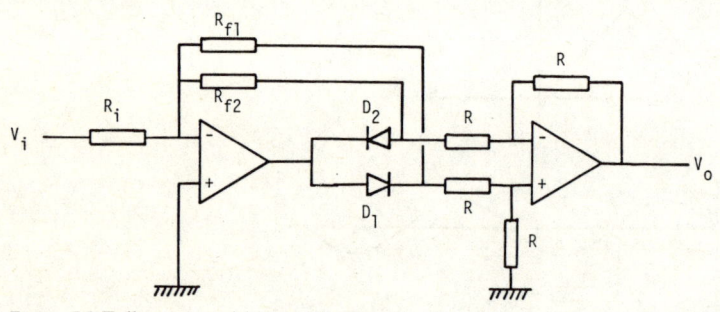

*Figure 7.8* Full-wave precision rectifier circuit

Negative values of $V_i$ cause feedback via $D_1$ and $R_{f1}$; positive half-cycles appear at the cathode of $D_1$. Similarly, positive inputs cause feedback via $D_2$ and $R_{f2}$ with negative half-cycles appearing at the anode of $D_2$. Notice that, in this configuration, a feedback path is provided for both positive and negative inputs and additional diodes to avoid saturation are not needed. These half-cycles are then summed differentially in order to provide the required output $V_o$. $R_{f1}$ and $R_{f2}$ will normally be of equal value and chosen to give the required gain. For simplicity, the differential amplifier has been arranged for unity gain but this is not essential.

For measurement applications, a ripple-free output from the rectifier is normally needed. This can be achieved by means of low pass filtering. It may also be required to block D.C. and very low frequency components in the input; this implies high pass filtering.

## 7.4 Arbitrary function generators

The circuits discussed in Sections 7.2 and 7.3 used diodes in order to realize certain specific simple non-linear functions. These techniques can readily be extended in order to realize any required function by means of a series of piece-wise linear approximations. Inevitably some error will be involved, but this can be made small by using a sufficiently large number of segments in the approximation. A detailed analysis of the errors involved is beyond the scope of this book; see, for example, References 7.2 and 7.3.

One possible approach to diode function generation is shown in basic form in Figure 7.9. Each diode is biased from a negative reference supply (which could be one of the operational amplifier power supplies if this is sufficiently stable) via a resistance $R_{Bj}$ (where $j$ takes a value between 1 and $n$). The bias arrangement is such that, for negative and small positive values of $V_i$, all diodes are reverse biased and $V_o$ is given by $-(R_f/R_i)V_i$ as usual. As $V_i$ becomes more positive,

The precision rectifier circuit may therefore be used in conjunction with an active filter configuration (see Chapter 6 and Reference 7.1).

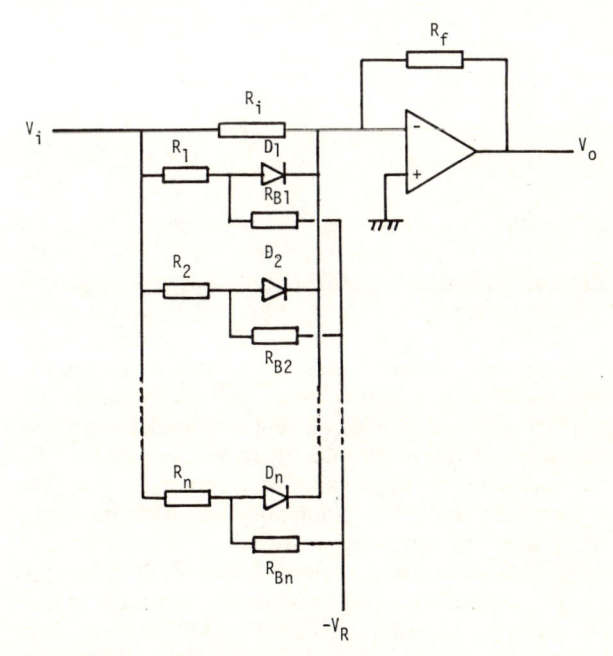

*Figure 7.9* Basic diode function generator

96  Non-linear circuits

the diodes will start to conduct. As this happens, the effective input resistance is reduced since $R_1, R_2$ etc. become effectively connected in parallel with $R_i$. The type of transfer characteristic that results is shown in Figure 7.10.

Since the cathode of each diode is connected to the virtual earth point, conduction will start for the $j$th diode $D_j$ when

$$-V_r + \left(\frac{R_{Bj}}{R_j + R_{Bj}}\right)(V_i + V_R) = 0$$

that is

$$V_i = V_R R_j / R_{Bj} \qquad (7.6)$$

(These values will of course be in error by about 0.5 volt due to the forward voltage drop of non-ideal diodes.)

Assuming that values are chosen such that $D_1$ conducts first, followed by $D_2$ and so on, the slopes shown in Figure 7.10 will be given by:

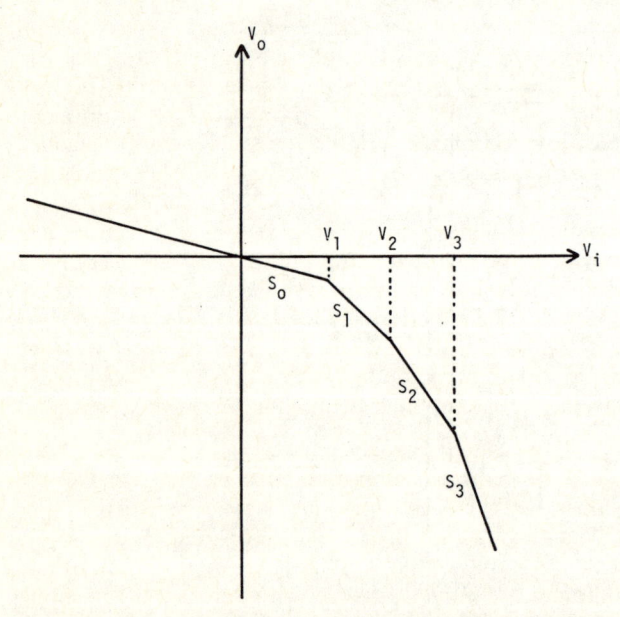

$$S_o = -\frac{R_f}{R_i}$$

Figure 7.10 Transfer characteristic of basic diode function generator

$$S_1 = -\frac{R_f}{R_i \text{ and } R_1 \text{ in parallel}}$$

$$= -R_f\left(\frac{1}{R_i} + \frac{1}{R_1}\right)$$

In general:

$$S_j = -R_f\left(\frac{1}{R_i} + \frac{1}{R_1} + \frac{1}{R_2} + \cdots \frac{1}{R_j}\right) \tag{7.7}$$

The circuit described has two major restrictions:

(a) it operates for only one polarity of input,
(b) additional input resistors can only increase the slope of the characteristic; that is $S_1 > S_o$, $S_2 > S_1$ etc. This is a 'monotonic' (single-slope) characteristic.

The restriction on input polarity can be removed by introducing a second, parallel connected, input network of the kind shown in Figure 7.9. The polarities of the diodes and the reference voltage should be reversed; the characteristic, however, will still be monotonic with an increasing slope.

Monotonic function generators with a decreasing slope can be realized by connecting a similar diode resistor network in the feedback path of the amplifier (with the diodes connected to the summing junction and the 'input resistors' connected to the amplifier output). Subsequent parallel connection of feedback resistances causes a decrease in the slope of the characteristic. Further details are given in Reference 7.4.

Monotonic function generators, as described in this section, are capable of providing the characteristics that are normally required such as square law, cube law, and sine wave shaping. More complex circuits have been devised which can generate an arbitrary sequence of slopes; see, for example, Reference 6.10.

Accurate design of a general diode function generator is complex; program P(7.3) has been simplified by making several assumptions. These are:

(1) The function to be realized is monotonic with an increasing slope. Decreasing slopes could be accommodated by a network in the feedback path but this has not been included in the program.
(2) The full scale input and output voltages are equal to each other.
(3) The breakpoints are equally spaced with respect to the input voltage range. Improved accuracy can be obtained if more

98  Non-linear circuits

```
10 REM PROGRAM P(7.3)
20 REM SIMPLE DIODE FUNCTION GENERATOR
30 REM THE REQUIRED FUNCTION MUST BE MONOTONIC
40 REM WITH AN INCREASING SLOPE
50 REM TO CHANGE THE FUNCTION, CHANGE LINE 60
60 DEF FNX(X)=X↑2
70 DIM BP(20),V0(20),S(20)
80 DIM R(20),RB(20),K(20)
90 PRINT ""
100 PRINT "NUMBER OF BREAKPOINTS"
110 INPUT NB
120 PRINT "FULL SCALE VOLTAGE"
130 INPUT FS
140 PRINT "REFERENCE VOLTAGE"
150 INPUT VR
160 PRINT "FEEDBACK RESISTANCE (KILOHMS)"
170 INPUT RF
180 REM CALCULATE BREAKPOINTS AND
190 REM THE CORRESPONDING OUTPUT VOLTAGES
200 REM AND SLOPES
210 FOR I = 1 TO (NB+1)
220 BP(I)=I*(FS/(NB+1))
230 VI = BP(I)/FS
240 V0(I) = FS*FNX(VI)
250 S(I-1)=(V0(I)-V0(I-1))/(FS/(NB+1))
260 NEXT I
270 REM CALCULATE RI
280 RI = RF/S(0)
290 REM ROUTINE TO CALCULATE INPUT RESISTANCES
300 R(1) = 1/((S(1)/RF)-(1/RI))
310 K(1) = (1/RI) + (1/R(1))
320 FOR J = 2 TO NB
330 R(J) = 1/((S(J)/RF) - K(J-1))
340 K(J) = K(J-1)+ (1/R(J))
350 NEXT J
360 REM CALCULATE BIAS RESISTANCES
370 FOR J = 1 TO NB
380 RB(J) = VR*R(J)/BP(J)
390 NEXT J
400 PRINT "DO YOU REQUIRE THE BREAKPOINTS (BP),"
410 PRINT "THE CORRESPONDING OUTPUT VOLTAGES (VO),"
420 PRINT " SLOPES (S), INPUT RESISTANCES (IR)"
430 PRINT "OR BIAS RESISTANCES (BR)"
440 INPUT A$
450 IF A$ = "BP" THEN 510
460 IF A$ = "VO" THEN 580
470 IF A$ = "S" THEN 650
480 IF A$ = "IR" THEN 720
490 IF A$ = "BR" THEN 810
500 GOTO 400
510 REM PRINT BREAKPOINTS
520 PRINT ""
530 FOR I = 1 TO (NB+1)
540 PRINT "BREAKPOINT"I"="BP(I)"VOLTS"
550 NEXT I
560 PRINT:PRINT
570 GOTO 400
580 REM PRINT OUTPUT VOLTAGES
```

```
590 PRINT ""
600 FOR I = 1 TO (NB+1)
610 PRINT "OUTPUT AT BP("I")="V0(I)"VOLTS"
620 NEXT I
630 PRINT:PRINT
640 GOTO 400
650 REM PRINT SLOPES
660 PRINT ""
670 FOR I = 1 TO (NB+1)
680 PRINT "S"I-1"="S(I-1)
690 NEXT I
700 PRINT:PRINT
710 GOTO 400
720 REM PRINT INPUT RESISTANCES
730 PRINT ""
740 PRINT "RF="RF"KILOHMS"
750 PRINT "RI="RI"KILOHMS"
760 FOR I = 1 TO NB
770 PRINT "R"I"="R(I)"KILOHMS"
780 NEXT I
790 PRINT:PRINT
800 GOTO 400
810 REM PRINT BIAS RESISTANCES
820 PRINT ""
830 FOR I = 1 TO NB
840 PRINT "RB"I"="RB(I)"KILOHMS"
850 NEXT I
860 PRINT:PRINT
870 GOTO 400
880 END
```

breakpoints are used in regions where the required function changes rapidly (Refs 7.2 and 7.3).

(4) The diodes are ideal. In practice, the diodes will have a forward voltage drop (about 0.6 volt) and a finite 'on' resistance. The former could be allowed for by a suitable modification to Equation (7.6) and the latter by ensuring that the input resistances used are large compared with the diode resistances. Notice that if a breakpoint of less than about 0.6 volt is required it would, in practice, be necessary to introduce a diode and reference supply of opposite polarity.

(5) The required function must be finite over the whole input range. A tangent function, for example, could not be generated over the range 0 to 90°.

Within these limitations, program P(7.3) gives a good 'feel' for the component values required in a typical function generator. The required function is defined directly in line 60. The arrays for the breakpoints (BP), corresponding output voltages ((VO), slopes (S), input resistances (R) and bias resistances (RB) are dimensioned in lines 70 and 80. K is an intermediate value and used in the calculation of input resistances. An array size of 20 should be adequate; it can be changed if required. The parameters requested in lines 100 to 170 should be self-explanatory.

The number of segments in the characteristic will be one more than the number of breakpoints (line 210) and the corresponding, equally spaced, breakpoints are determined in line 220. The input voltage, at the breakpoint, is normalized with respect to full scale (line 230) and the corresponding required output voltage is determined (line 240). The slope of each segment will be the difference between the two appropriate output voltages divided by the input voltage change (line 250).

Calculation of $R_i$ from the feedback resistance and initial slope is straightforward (line 280), but determination of the required input resistances (from Equation (7.7)) is slightly more difficult. For the second slope $(S_1)$:

$$S_1 = R_f\left(\frac{1}{R_i} + \frac{1}{R_1}\right) \quad \text{(ignoring the negative sign)}$$

Therefore

$$R_1 = \left(\frac{S_1}{R_f} - \frac{1}{R_i}\right)^{-1}$$

as calculated in line 300.

For each subsequent input resistance an additional term is required in (7.7) and $R_j$ cannot be calculated until $R_{j-1}$ has been determined. In the program this is resolved as follows:

Let

$$\frac{1}{R_i} + \frac{1}{R_1} + \frac{1}{R_2} + \cdots \frac{1}{R_j} = K_j$$

and

$$K_j = K_{j-1} + \frac{1}{R_j}$$

From (7.7)

$$S_j = R_f K_j \quad \text{(ignoring the sign again)}$$
$$= R_f\left(K_{j-1} + \frac{1}{R_j}\right)$$

Therefore

$$R_j = \left(\frac{S_j}{R_f} - K_{j-1}\right)^{-1}$$

which is evaluated in line 330, and $K_j$ is updated in line 340.

Having obtained the input resistances, the bias resistances are easily calculated using Equation (7.6) (line 380).

The information generated by this program exceeds the display capabilities of most small computers. Therefore a choice is provided (lines 400 to 440) followed by a jump to the appropriate PRINT routine (lines 450 to 490). An invalid response causes a return to line 400, as does the termination of each PRINT routine. Consequently, the output data may be accessed in any order and repeated if required. (The program must be terminated with the BREAK key or an equivalent function.)

In the specimen run each of the output routines is called in turn but this of course is not necessary. For example, the slopes are useful for checking purposes but are not needed when all is well. The resistance values would be needed during construction and input/output voltages for test purposes.

Notice that the input resistance list also contains the feedback resistance (as specified by the user) and the input resistance as calculated in the program.

*Specimen runs*
```
NUMBER OF BREAKPOINTS
? 8
FULL SCALE VOLTAGE
? 10
REFERENCE VOLTAGE
? 12
FEEDBACK RESISTANCE (KILOHMS)
? 47
DO YOU REQUIRE THE BREAKPOINTS (BP),
THE CORRESPONDING OUTPUT VOLTAGES (VO),
 SLOPES (S), INPUT RESISTANCES (IR)
OR BIAS RESISTANCES (BR)
?BP
BREAKPOINT 1 = 1.11111111 VOLTS
BREAKPOINT 2 = 2.22222222 VOLTS
BREAKPOINT 3 = 3.33333333 VOLTS
BREAKPOINT 4 = 4.44444445 VOLTS
BREAKPOINT 5 = 5.55555556 VOLTS
BREAKPOINT 6 = 6.66666667 VOLTS
BREAKPOINT 7 = 7.77777778 VOLTS
BREAKPOINT 8 = 8.88888889 VOLTS
BREAKPOINT 9 = 10 VOLTS
DO YOU REQUIRE THE BREAKPOINTS (BP),
```

## 102 Non-linear circuits

```
THE CORRESPONDING OUTPUT VOLTAGES (VO),
 SLOPES (S), INPUT RESISTANCES (IR)
OR BIAS RESISTANCES (BR)
?VO
OUTPUT AT BP( 1 )= .12345679 VOLTS
OUTPUT AT BP( 2 )= .493827161 VOLTS
OUTPUT AT BP( 3 )= 1.11111111 VOLTS
OUTPUT AT BP( 4 )= 1.97530864 VOLTS
OUTPUT AT BP( 5 )= 3.08641975 VOLTS
OUTPUT AT BP( 6 )= 4.44444445 VOLTS
OUTPUT AT BP( 7 )= 6.04938272 VOLTS
OUTPUT AT BP( 8 )= 7.90123458 VOLTS
OUTPUT AT BP( 9 )= 10 VOLTS
DO YOU REQUIRE THE BREAKPOINTS (BP),
THE CORRESPONDING OUTPUT VOLTAGES (VO),
 SLOPES (S), INPUT RESISTANCES (IR)
OR BIAS RESISTANCES (BR)
?S
S 0 = .111111111
S 1 = .333333333
S 2 = .555555555
S 3 = .777777779
S 4 = .999999999
S 5 = 1.22222222
S 6 = 1.44444445
S 7 = 1.66666667
S 8 = 1.88888889
DO YOU REQUIRE THE BREAKPOINTS (BP),
THE CORRESPONDING OUTPUT VOLTAGES (VO),
 SLOPES (S), INPUT RESISTANCES (IR)
OR BIAS RESISTANCES (BR)
?IR
RF= 47 KILOHMS
RI= 423 KILOHMS
R 1 = 211.5 KILOHMS
R 2 = 211.5 KILOHMS
R 3 = 211.499999 KILOHMS
R 4 = 211.500002 KILOHMS
R 5 = 211.499999 KILOHMS
R 6 = 211.499994 KILOHMS
R 7 = 211.500006 KILOHMS
R 8 = 211.500003 KILOHMS
DO YOU REQUIRE THE BREAKPOINTS (BP),
THE CORRESPONDING OUTPUT VOLTAGES (VO),
 SLOPES (S), INPUT RESISTANCES (IR)
OR BIAS RESISTANCES (BR)
?BR
RB 1 = 2284.2 KILOHMS
RB 2 = 1142.1 KILOHMS
RB 3 = 761.399995 KILOHMS
RB 4 = 571.050007 KILOHMS
RB 5 = 456.839998 KILOHMS
RB 6 = 380.69999 KILOHMS
RB 7 = 326.314295 KILOHMS
RB 8 = 285.525004 KILOHMS
DO YOU REQUIRE THE BREAKPOINTS (BP),
THE CORRESPONDING OUTPUT VOLTAGES (VO),
 SLOPES (S), INPUT RESISTANCES (IR)
OR BIAS RESISTANCES (BR)
```

## 7.5 Logarithmic amplifiers

In some special cases, the biased diode networks, discussed in the previous section, can be replaced with a single non-linear component. In particular, the exponential nature of the voltage/current relationship for a p–n semiconductor junction has made simple and effective logarithmic (and antilogarithmic) amplifiers possible.

The logarithmic relationship between amplifier input and output has a wide range of important applications. These include:

(a) dynamic range compression and expansion to modify peak to average power ratios in communication systems;
(b) generation of products, quotients and powers by adding, subtracting and scaling logarithms respectively; and
(c) display of a measured quantity directly in decibels, for example by a spectrum analyser.

The required p–n junction is conveniently part of a transistor, as shown in Figure 7.11. For an n–p–n transistor, the Ebers–Moll equations (Ref 7.5) give

$$I_C = \alpha I_{ES}(e^{-qV_e/kT} - 1) - I_{CS}(e^{-qV_C/kT} - 1) \qquad (7.8)$$

where $I_C$ is the collector current,
$\alpha$ is the common base current gain,
$I_{ES}$ and $I_{CS}$ are the emitter and collector saturation currents respectively,
$q$ is the charge on the electron,
$V_e$ and $V_C$ are the emitter and collector voltages respectively,
$k$ is Boltzmann's constant,
and $T$ is the absolute temperature.

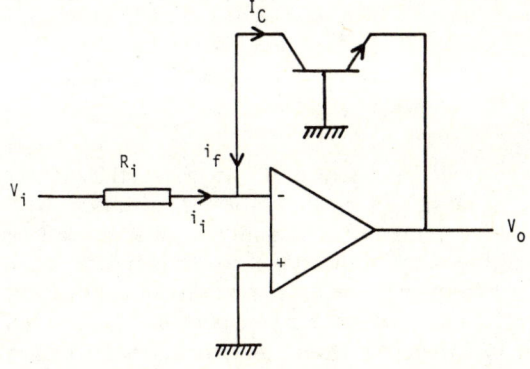

Figure 7.11 Basic logarithmic amplifier

104  Non-linear circuits

In Figure 7.11, the collector is connected to the virtual earth point so, by feedback action, $V_C$ is close to zero and the second term in Equation (7.8) vanishes. Also, from Figure 7.11 $V_e = V_o$ and $i_f = -I_C$. Summing currents in the usual way:

$$i_i + i_f = 0$$

$$\frac{V_i}{R_i} = \alpha I_{ES}(e^{-qV_o/kT} - 1)$$

Now $I_{ES}$ is very small (of the order of $10^{-13}$ amp) so, for realistic input and feedback current, the exponential term will be large compared with 1 and the equation reduces to

$$\frac{V_i}{R_i} = \alpha I_{ES}\, e^{-qV_o/kT}$$

Taking natural logarithms of both sides

$$\ln \frac{V_i}{R_i} = -\ln \alpha I_{ES}(qV_o/kT)$$

Therefore

$$V_o = -\frac{kT}{q} \ln \frac{V_i}{R_i \alpha I_{ES}} \tag{7.9}$$

$$= -\frac{kT}{q} \ln \frac{V_i}{V_R}$$

where $V_R$ is a reference voltage equal to $R_i \alpha I_{ES}$.

The output, therefore, is proportional to the logarithm of the input as required. The exponential function necessitated the use of natural logarithms but base 10 can be accommodated by suitable scaling since

$$\log_{10} x = \ln x \log_{10} e$$
$$= 0.4343 \ln x$$

Equation (7.9) shows two points of difficulty. First, the emitter saturation current $I_{ES}$, although small, varies significantly with temperature and between transistors. Secondly, even if $I_{ES}$ can be compensated, $V_o$ is still directly proportional to temperature.

The first effect is overcome by using a second circuit of the form shown in Figure 7.11. This is fed from a constant (but adjustable) supply $V_C$. Its output $V_{OC}$, therefore, from equation (7.9) will be given by

$$V_{OC} = -\frac{kT}{q}\ln\frac{V_C}{R_i\alpha I_{ES}}$$

If the two amplifier outputs are combined in a differential amplifier the result is

$$V'_o = V_o - V_{OC} = -\frac{kT}{q}\ln\left(\frac{V_i}{R_i\alpha I_{ES}} - \ln\frac{V_C}{R_i\alpha I_{ES}}\right)$$

If the same value of $R_i$ is used for each amplifier and the values of $\alpha$ and $I_{ES}$ are similar (ensured by fabrication on the same integrated circuit wafer) this expression reduces to

$$V'_o = -\frac{kT}{q}\ln\frac{V_i}{V_C}$$

and $V_C$ can conveniently be used for calibration.

Unfortunately, the direct dependence of the output on temperature remains. Its effect can be alleviated by means of a voltage follower stage (Figure 2.6(a)) in which the feedback network includes temperature-sensitive elements to provide the required variation of gain with temperature.

It can be seen that it is not easy to design a logarithmic amplifier whose accuracy will be maintained over a wide range of input voltages and temperatures. From a user's point of view the solution is to use a well-established integrated circuit module (such as the Analog Devices 755). This device is also capable of inverse connection in order to provide an antilogarithm function. A good treatment of amplifiers of this type is provided in Reference 7.6.

## 7.6 Exercises

**(7.1)** Extend program P(7.1) so that limit values may be specified with respect to the input.

**(7.2)** Extend program P(7.1) to allow for non-zero forward voltage drop in the feedback diodes.

**(7.3)** Derive Equations (7.4) and (7.5), which relate to Figure 7.5

**(7.4)** Extend program P(7.3) to cater for inputs of both polarities.

**(7.5)** Modify program P(7.3) to cater for diode feedback networks which provide a decreasing slope of the characteristic.

**(7.6)** Extend program P(7.3) to calculate the maximum error introduced by the piece-wise linear approximation.

**(7.7)** Modify program P(7.3) so that the user can select the breakpoint locations.

**(7.8)** Extend program P(7.3) to allow for typical values of diode forward voltage drop.

## 7.7 References

7.1 GARRETT, P. H. *Analog I/O design*. Reston Publishing 1981, pp 47–9
7.2 RITCHIE, C. C. and YOUNG, R. W. The design of biased diode function generators. *Electronic Engineering* Vol 31, p. 347, 1959
7.3 REAM, N. Approximation errors in diode function generators. *Journal of Electronics and Control* Vol 7, p. 83, 1959
7.4 JACOB, J. M. *Applications and Design with Analog Integrated Circuits*. Reston 1982, pp 409–21
7.5 EBERS, J. J. and MOLL, J. L. The large signal behaviour of junction transistors. *Proc IRE* Vol 42, pp 1761–72, 1954
7.6 As 7.4, pp 427–42

Appendix
# Preferred component values

In order to rationalize production, many electronic components are manufactured in a range of 'preferred' values. The increments chosen define the maximum error which can occur when using a preferred value in place of that determined by a design calculation. For example, the widely used E12 series (as specified by lines 60 to 180 of program P(A.1)) has increments of approximately 20 per cent, which means that an arbitrary value can be approximated to within 10 per cent or better.

Program P(A.1) was written primarily for resistance values, hence the range test in line 210 for values between 0.1 $\Omega$ and 100 M$\Omega$ (the normally available commercial range). With appropriate modification to this line the program is equally suitable for other types of component.

Of course, the range specified between lines 60 and 180 represents only one decade and the required value could be one of these multiplied by an appropriate power of 10. The required value (R) is therefore shifted into range by lines 220 to 250 which perform repeated multiplication or division as required. M records the multiplier value selected.

The best match is then obtained in lines 270 to 310. I is increased (line 280) until E(I) exceeds R. The difference between these two values (the error) is calculated (as EH) in line 300. However, it could well be that the immediately lower preferred value, although less than R, represents a smaller error so this (EL) is calculated in line 290. These two errors are then compared in line 310 and I is reduced by one if appropriate. Notice that, since relative rather than absolute errors are normally important, it is EH/R and EL/R which are compared.

Finally, the chosen value is printed (line 330), not forgetting the multiplier M, and the resulting error is calculated and printed in line 350.

The demonstration run shows typical operation of the program. In particular, notice the jump from R = 299 to R = 300 where the

## Appendix

```
10 REM PROGRAM P(A.1)
20 REM PROGRAM TO FIND NEAREST PREFERRED VALUE
30 PRINT""
40 REM SET UP ARRAY OF PREFERRED VALUES
50 DIM E(13)
60 E(1)=10
70 E(2)=12
80 E(3)=15
90 E(4)=18
100 E(5)=22
110 E(6)=27
120 E(7)=33
130 E(8)=39
140 E(9)=47
150 E(10)=56
160 E(11)=68
170 E(12)=82
180 E(13)=100
190 PRINT "REQUIRED RESISTANCE IN OHMS"
200 INPUT R
210 IF R<0.1 OR R>1E08 THEN PRINT"OUT OF RANGE":GOTO 190
220 REM SHIFT R INTO RANGE
230 M=1
240 IF R<10 THEN R=R*10:M=M/10:GOTO240
250 IF R>=100 THEN R=R/10:M=M*10:GOTO250
260 I=1
270 REM FIND NEAREST E(I)
280 IF E(I)<R THEN I=I+1:GOTO280
290 EL=R-E(I-1)
300 EH=E(I)-R
310 IF EL/R<EH/R  THEN I=I-1
320 PRINT
330 PRINT E(I)*M"IS THE NEAREST PREFERRED VALUE"
340 PRINT
350 PRINT"ERROR="(E(I)-R)*100/R"PERCENT"
360 PRINT
370 GOTO 190
```

## Specimen run

```
REQUIRED RESISTANCE IN OHMS
? .01
OUT OF RANGE
REQUIRED RESISTANCE IN OHMS
? .11
 .12 IS THE NEAREST PREFERRED VALUE
ERROR= 9.09090909 PERCENT
REQUIRED RESISTANCE IN OHMS
? 290
 270 IS THE NEAREST PREFERRED VALUE
ERROR=-6.89655173 PERCENT
REQUIRED RESISTANCE IN OHMS
? 295
 270 IS THE NEAREST PREFERRED VALUE
ERROR=-8.47457628 PERCENT
REQUIRED RESISTANCE IN OHMS
? 299
```

```
 270 IS THE NEAREST PREFERRED VALUE
ERROR=-9.69899665 PERCENT
REQUIRED RESISTANCE IN OHMS
? 300
 330 IS THE NEAREST PREFERRED VALUE
ERROR= 10 PERCENT
REQUIRED RESISTANCE IN OHMS
```

program changes from choosing the immediately lower to the immediately higher preferred value.

By changing the values specified between lines 50 and 180, the program can easily be modified to accept other series of preferred values.

The program can be incorporated, as a subroutine, into many of the design programs given in this book. In this case the error calculation (line 350) could be extended to include the required parameters of the circuit (gain, break frequency, etc.).

A program of this kind would be particularly useful as an aid to minimizing errors while retaining the use of preferred value components.

# Index

Active filter, 57, 95
Amplifier,
 buffer, 13, 57, 80, 91
 differential, 14, 77, 93, 105
Amplifier gain, 5
 instrumentation, 17
 logarithmic, 103
Amplitude stabilization, 53
Analogue computer, 77
Antilogarithm function, 105
Arbitrary function generator, 95
Array, 1, 33, 99
Attenuator, 42

Band pass filter, 63, 73
Band rejection filter, 79
Bandwidth,
 closed loop, 28
 full power, 29, 31
Bessel filter, 66
Bias current, 37, 40, 43, 63
Bias resistance, 101
Blocking capacitor, 44
Bode plot, 27, 63
Break frequency, 27, 61, 63
BREAK key, 101
Breakpoint, 97
Bubble sort routine, 33
Buffer amplifier, 13, 57, 80, 91
Butterworth filter, 66

Centre frequency, 73
Characteristic, transfer, 5
Chebyshev filter, 66
Closed loop operation, 6, 28
CLS command, 1
CMRR, 17
Common mode gain, 17, 21
Common mode operation, 16
Common mode rejection, 16
Common zero, 5, 7

Comparator amplifier, 6, 49
Compensating resistor, 37
Compensation, 24

Damping coefficient, 65, 70, 74
Damping factor, 64
D.C. blocking, 44
D.C. gain, 25
Decade, 26
Decibel, 26, 103
Difference current, 37
Difference signal, 14
Differential amplifier, 14, 77, 93, 105
Differential mode, 14, 16
DIMension statement, 1, 33
Diode bridge, 91
Diode characteristic, 86, 90, 96, 99
Direct mode, 16
Distortion, 31, 47, 54
Dollar sign, 4
Dynamic range, 103

E12 series, 107
Earth, virtual, 8
Ebers–Moll equations, 103
Effective feedback resistance, 43
Effective input resistance, 13
Equal component filter, 70
Error,
 offset, 9, 37
 output, 44
External compensation, 24

Feedback, 5
Feedback limiter, 85
Feedback resistance, effective, 43
Feedback resistor, 6
FET input amplifier, 39
Filter,
 active, 57
 band pass, 63, 73

111

Filter, *continued*
   band rejection, 79
   Bessel, 66
   Butterworth, 66
   Chebyshev, 66
   equal component, 70
   first order, 60
   high pass, 72, 79
   low pass, 26, 64, 78, 94
   notch, 81
   order of, 58
   passive, 57
   Sallen and Key, 67
   second order, 63, 72, 78
   state variable, 77
   unity gain, 67
First order active filter, 60
Follower, voltage, 11, 13, 105
FOR...NEXT, 3
Frequency,
   break, 27, 61
   natural, 64, 70
   oscillation, 54
Frequency response, 24
Full power bandwidth, 29, 31
Full wave rectification, 93
Function generator, 95

Gain, amplifier, 5
Gain–bandwidth product, 27, 33
Gain,
   common mode, 17, 21
   D.C., 25
   differential mode, 17
   integrator, 48
   loop, 54
   open loop, 5, 25
GET statement, 3
GOTO statement, 4

Half power point, 27
Half wave rectification, 92
High pass characteristic, 62, 79
High pass filter, 72
Hysteresis, 49

Ideal operation, 5
Ideal voltage source, 9
IF...THEN, 3
INKEY command, 3
Input bias current, 37, 43
Input offset current, 38
Input resistance, 9, 13, 16
Input resistor, 6

INPUT statement, 3
Instrumentation amplifier, 17
Integrated circuit, 105
Integration, 48
Internal compensation, 24
Inverting input, 6
Inverting mode, 6

Junction, summing, 8

Laplace operator, 64
Large signal operation, 29
LET command, 3
Limiter circuit, 84, 89
Logarithm, 18, 104
Logarithmic amplifier, 103
Loop, 3
Loop gain, 54
Low pass filter, 24, 64, 78, 94

Mode,
   common, 16
   differential, 14
   direct, 16
   non-inverting, 11
   series, 16
Monotonic characteristic, 97

Natural frequency, 64, 70
NEXT, 3
Non-inverting input, 6
Non-inverting mode, 11
Non-linear circuits, 84, 103
Notch filter, 80

Octave, 26
Offset, total, 40
Offset error, 9, 37
Offset voltage, 37, 40
Open loop bandwidth, 24
Open loop gain, 5, 24, 25
Open loop operation, 6
Operation,
   ideal, 5
   large signal, 29
   small signal, 30
Order of filter, 58
Oscillator, since wave, 51
Output resistance, 9, 16
Output saturation voltage, 5

Parameter sensitivity, 75, 79
Phase, 25, 62, 63, 72
Piece-wise linear approximation, 95

Power supply, 5
Precision limiter, 89
Precision rectifier, 92
Preferred value, 107
PRINT statement, 1, 101

Q factor, 73, 80
Quotation marks, 3

Ramp generation, 47
Rectification, 84, 89, 92
Rejection, common mode, 16
REMark, 1
Resistance,
  input, 9, 16
  output, 9, 16
RETURN key, 2
Rise time, 28, 31
Roll-off, 26
RUN command, 4

Sallen and Key filter, 67
Saturation, 5, 49, 91, 104
Sawtooth wave generator, 51
Scaling amplifier, 6, 8, 13
Second order active filter, 63, 72, 78
Series mode, 16
Sine wave oscillator, 51
Slew rate, 29
Small signal operation, 30
Soft limiting, 87

Sort routine, 33
Square wave generator, 50
State variable filter, 77
String variable, 4
Subroutine, 109
Subtraction, 8
Summing amplifier, 6, 13
Summing junction, 8
Supply voltage, 40

T network, 42, 80
TAB command, 3
Temperature effects, 40
THEN, 3
Transfer characteristic, 5, 96
Transfer function, 58
Triangular wave generator, 50
Trimming potentiometer, 40

Unity gain buffer, 13
Unity gain filter, 67

Virtual earth, 8
Voltage follower, 11, 13, 105
Voltage source, ideal, 9

Waveform generation, 47
Wien bridge, 52

Zener diode, 92
Zero, common, 5, 7